The ORIGIN of the THIRD FAMILY

World Scientific Series in 20th Century Physics

Published

For information on Vols. 1–24, please visit http://www.worldscientific.com/series/wsscp

World Scientific Series in 20th Century Physics – Vol. 20

The ORIGIN *of the* THIRD FAMILY

IN HONOUR OF A. ZICHICHI ON THE XXX ANNIVERSARY OF THE PROPOSAL TO SEARCH FOR THE THIRD LEPTON AT ADONE

**C.S. Wu, T.D. Lee, N. Cabibbo,
V.F. Weisskopf, S.C.C. Ting, C. Villi,
M. Conversi, A. Petermann,
B.H. Wiik and G. Wolf**

Edited by

O. Barnabei, L. Maiani, R.A. Ricci and F. Roversi Monaco
Academy of Sciences • INFN • SIF • Bologna University

World Scientific
Singapore • New Jersey • London • Hong Kong

Published by

World Scientific Publishing Co. Pte. Ltd.

5 Toh Tuck Link, Singapore 596224

USA office: 27 Warren Street, Suite 401-402, Hackensack, NJ 07601

UK office: 57 Shelton Street, Covent Garden, London WC2H 9HE

Library of Congress Cataloging-in-Publication Data
The origin of the third family / C.S. Wu ... [et al.] : editors, O. Barnabei,
    L. Maiani, R.A. Ricci and F. Roversi Monaco.
        p. cm. -- (World Scientific series in 20th century physics; vol. 20)
    Includes bibliographical references.
    ISBN-13 978-981-02-3163-7 -- ISBN-10 981-02-3163-6
    1. Heavy leptons (Nuclear physics) I. Wu, C.S. (Chien-shiung), 1912–
II. Barnabei, O. (Ottavio) III. Maiani, L. (Luciano)
IV. Ricci, R.A. (Renato Angelo) V. Monaco, R. Roversi (Fabio Roversi) VI. Series.
QC793.5.H32075 1998
539.7'2ll--dc21   98-20205
                                                                                        CIP

British Library Cataloguing-in-Publication Data
A catalogue record for this book is available from the British Library.

First edition jointly published in 1997 by Accademia Delle Scienze Bologna, Alma Mater Studiorum, lstituto
Nazionale di Fisica Nucleare and Societa ltaliana di Fisica.

The most favourable mechanism for the production of a heavy lepton HL is

$$(1) \qquad \qquad e^+e^- \to HL + \overline{HL} ,$$

which, in the one-photon approximation, is described by the Feynman diagram

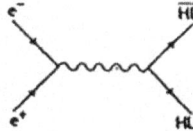

By analogy with muon decay, we may expect the predominant decay channel to be

$$(2) \qquad \qquad HL^{\pm} \begin{cases} \to e^{\pm} \nu_e \nu_H , \\ \to \mu^{\pm} \nu_\mu \nu_H \end{cases}$$

where ν_H indicates the neutrino proper to the heavy lepton HL. If this is the case, and the two modes are, as expected, equally probable, then there is $\sim 50\%$ probability that the final state from reaction (1) consists of a noncollinear $e^{\pm}\mu^{\mp}$ pair.

From: *"Limits on the Electromagnetic Production of Heavy Leptons"*, Lettere al Nuovo Cimento, 4, 1156 (1970).

The introduction from the paper published by the BCF group, where the first limit on the HL search at ADONE was reported.

This book is dedicated to three Founding Fathers of European Physics and CERN: Patrick Maynard Stuart BLACKETT, Enrico FERMI and Isidor Isaac RABI.

Chien Shiung Wu

CONTENTS

FOREWORD

The interest raised by this volume and its success have prompted the publication of this second edition. It enables interested readers to have at their disposal a source of basic material on the Origin of the Third Family of fundamental particles.

On 20 March 1967 a group of physicists from the University of Bologna and the Italian Institute for Nuclear Physics (INFN), led by A. Zichichi, published a proposal to search for a heavy lepton using the new INFN (e^+e^-) collider to be built at Frascati. The proposal, whose key pages are reproduced herein on the 30th anniversary of their publication, was the consequence of many years of work started at CERN where, in addition to the original idea of searching for a heavy lepton carrying its own leptonic number, new technologies were invented in order to allow the detection of a signal whose identification against the high background of hadronic processes was extremely difficult.

More than ten years of work by A. Zichichi, together with his students and his collaborators, have paved the way for the discovery of the Third Family of fundamental particles. The INFN, the University of Bologna, its Academy of Sciences and the Italian Physical Society (SIF) are proud to edit a volume where a group of eminent physicists, with their authority, unequivocally establishes the origin of the Third Family, a milestone of the Standard Model.

This volume is a joint publication, promoted by the Bologna Academy of Sciences and the Alma Mater Studiorum (University of Bologna), together with the Italian Institute for Nuclear Physics (INFN) and the Italian Physical Society (SIF). We thank the Director of the Physics Department of the University of Bologna, Professor Attilio Forino, the Director of INFN-Bologna, Professor Paolo Giusti, Professor Maurizio Basile and Professor Luisa Cifarelli, for their collaboration in the preparation of the volume.

Ottavio Barnabei
President of the Bologna Academy of Sciences

Luciano Maiani
President of the Italian Institute for Nuclear Physics

Renato Angelo Ricci
President of the Italian Physical Society

Fabio Roversi Monaco
Rector of the University of Bologna

Bologna, Padua, Rome, 20 March 1998

20 March 1967: The INFN-Bologna Proposal to Search for Heavy Leptons
Reproduction of Key Pages

Comitato Nazionale per L'Energia Nucleare
ISTITUTO NAZIONALE DI FISICA NUCLEARE

Sezione di Bologna
67/1

INFN/AE-67/3
20 Marzo 1967

M. Bernardini, D. Bollini, E. Fiorentino, F. Mainardi, T. Massam, L. Monari, F. Palmonari and A. Zichichi (Bologna-Cern-Frascati collaboration) : A PROPOSAL TO SEARCH FOR LEPTONIC QUARKS AND HEAVY LEPTONS PRODUCED BY ADONE. -

Reparto Tipografico
dei Laboratori Nazionali di Frascati

6.

In Fig. 4 we have plotted the differential cross-section at E = = 1.5 GeV for the above mentioned process.

The $\gamma\gamma$-rate is less than two orders of magnitude greater than the $\bar{q}$ q-rate, over the acceptance range of our telescope. As our pulse height analysis is in coincidence with a pair of thin counters the combined probability of a trigger is obviously negligible.

The details concerning the apparatus and the associated electronic logic are given in § 4.

3. - SEARCH FOR HEAVY LEPTONS -

If heavy leptons exist would they have been detected?

In order to attempt to answer this question we need to make few remarks: if we assume universality also for the coupling of this new heavy lepton to the known leptons, it turns out that the lifetime of a heavy lepton with 1 GeV mass would be of the order of 10^{-11} sec and could never have been detected as a decaying particle, but only as a resonance. Moreover the production of μ is copious only because of the fact that it is the decay product of a very commonly produced particle: the π. There is no equivalent mechanism for the production of a 1 GeV heavy lopton: in proton-machines they could only be produced in pairs via time-like photons, a process of which the low rate has already been discussed. Moreover the lack of stability of this particle is consistent with its apparent absence.

By studying the most favourable mechanisms which could produce the heavy leptons we reach the following conclusion. If in the process

$$e^+ e^- \rightarrow H_1^+ + H_1^-$$

we set at an energy E such that the ratio

$$\frac{E}{M_{H_1}} \simeq 1.2$$

as can be seen from Fig. 6 the cross-section is around 10^{-32} cm^2. Moreover the two produced H_1^+ and H_1^- are non relativistic and very slow in the laboratory-system, their $\gamma = E/M$ is in fact ~ 1.2. The most favoured decay channels, as far as we can say now, are probably

$$H_1^{\pm} \begin{array}{l} \longrightarrow e^{\pm} + \nu_e + \nu_H \\ \longrightarrow \mu^{\pm} + \nu_\mu + \nu_H \end{array}$$

7.

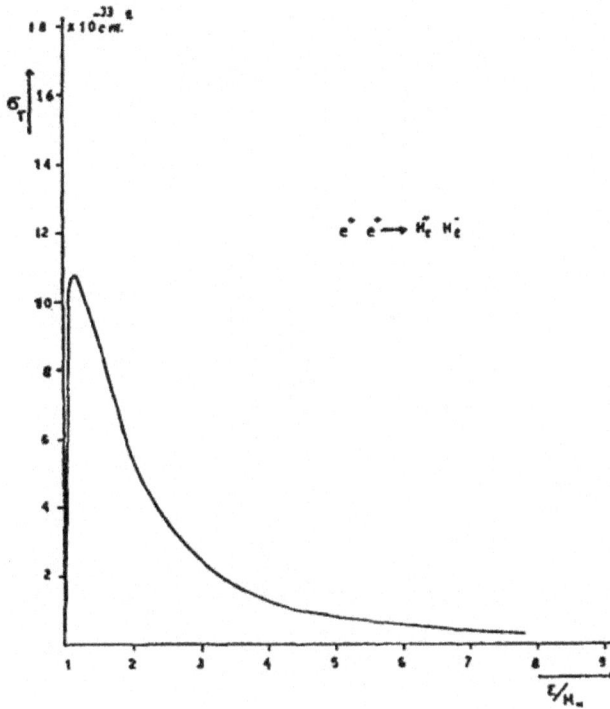

FIG. 6 - Total cross-section for production of heavy leptons versus E/M_{H_1}.

and therefore the decay angular distribution is nearly isotropic in the laboratory-system. This means that the probability of having the two charged light leptons going in the same hemisphere in the laboratory--system is ~50%. These events would be identified by what we call an "asymmetric" trigger, i.e. two particles in one telescope and nothing on the other side. Notice that this "asymmetric" trigger should be further characterized by the following typical lepton-pair distribution: e^+e^-: 25%; $\mu^+\mu^-$: 25%; $e^{\pm}\mu^{\mp}$: 50%.

The expected rate of these triggers is found to be ~1.5 events/ /hour for $E/M_{H_1} = 1.2$ and for the luminosity of Adone relative to E = = 1.2 GeV. This means that if a heavy lepton exists with $M_{H_1} = 1$ GeV we expect 36 events in one day of running with Adone at E = 1.2 GeV.

Chien Shiung Wu

THE ORIGIN OF THE THIRD FAMILY

Columbia University, New York, USA

THE ORIGIN OF THE THIRD FAMILY

Chien Shiung Wu

Most of my research has been on the properties of the lightest leptons: electrons, positrons and their associated neutrinos, especially through their weak interactions. I feel in some way justifiably proud to see that these light leptons have become members of a much larger family, and it gives me great pleasure to present this introduction concerning the extraordinary history of the origin and development of the Third Family of the leptons.

The origin of the Third Family of leptons and quarks is one of the most instructive examples, in our century, of how a new field can be opened: it is bound to remain as a landmark in the History of Physics. It took many decades to finally get an equal number of quarks and leptons grouped into two families (or generations)*. Two groups of decreasing-charges $(+2/3, 0, -1/3, -1)$, quarks and leptons, i.e. (u, v_e, d, e^-) and (c, v_μ, s, μ^-), appeared to be — after so many years — the basic structure of matter. But, as Shelley Glashow said at the 1994 International Conference on "The History of Original Ideas and Basic Discoveries in Particle Physics": *"That peculiar symmetry was short-lived and later the same year Marty Perl — inspired by Nino Zichichi's earlier but unsuccessful searches at lower energies — sought and found the tau lepton"*. This "peculiar symmetry" of two groups of decreasing-charge quarks and leptons was so "short-lived" because the earlier work of A. Zichichi towards the Third Family had already spanned over more than a decade.

During the late fifties the great majority of physicists were fully engaged in strong interactions. Nevertheless, a high precision experiment on the anomalous magnetic moment of the muon, performed at CERN, gave a clear indication that the muon was a particle totally deprived of any sort of interaction but the electromagnetic and the weak ones: its magnetic anomaly was, within five parts in a thousand, as expected from pure QED with radiative corrections included. But the muon mass was 200 times heavier than the electron. Once the existence of the muon as a heavy electron was well established, what was the next step to be?

A large variety of proposals were presented, but none of the type thought of and thoroughly investigated in all its consequences at CERN by A. Zichichi.

In fact, (i) the idea for the existence of a new heavy lepton (HL) with its own leptonic

---

(*) The term "generation" implies a sequential correlation between the three families. At present there is no evidence that the first family generates the second and this the third, or viceversa. The three families are not sequentially correlated, as the term "generation" would imply. This is why we prefer to adopt the term "family".

number and coupled to its own neutrino, (ii) how to search for it (acoplanar $e\mu$ pairs), (iii) the construction of the first experimental set-up able to reject the high background levels, (iv) the proof that the best source for HL pairs was not $(\bar{p}p)$ but (e^+e^-), and (v) the experimental evidence that the search for this new heavy lepton using (e^+e^-) colliders could very well be achieved: all these were Zichichi's work for more than a decade, first at CERN and later, with the advent of the (e^+e^-) collider ADONE, at Frascati. The search for a new lepton coupled to its own neutrino was not an obvious and simple matter to deal with. It is easy to do the right thing once you have the right ideas.

That all these matter were not trivial ones is proved by the fact that all the papers published before 1970, the date of the first Frascati results, never considered the idea of the heavy lepton proposed and searched for by A. Zichichi during more than a decade of experiments; the key point was its best signature, a new effect: the production, above some threshold energy, of acoplanar $(e^{\pm}\mu^{\mp})$ pairs produced by time-like photons.

This volume has two main components: reports and testimonies. Both will allow the reader to know how this new field of physics was opened, how it gave rise to new technological developments (now still of great value for electron and muon detection), and how much work was needed for the "peculiar symmetry" to be so "short-lived".

The first paper by T.D. Lee traces the history of the leptons, from the muon — originally interpreted as the Yukawa meson — up to the bold theoretical idea of a third leptonic number, which led to the discovery of $(e\mu)$ pairs in (e^+e^-) annihilation and therefore of the third lepton.

Nicola Cabibbo, Viki Weisskopf and Sam Ting recall the crucial years at CERN, from the $(g-2)$ of the muon, to the search for $(e\mu)$ pairs produced by time-like photons in $(\bar{p}p)$ annihilation. This search led to the construction of the most intense beam of (partially separated) antiprotons and to innovative technologies such as the pre-shower method (to improve electron detection) and muon punch-through studies (to improve muon detection), still now in use for new particle searches.

The basic steps which led this search from CERN to the new (e^+e^-) collider being implemented at Frascati are described by Claudio Villi, who was, in those years at the INFN, the physicist in charge of new experiments with the ADONE (e^+e^-) collider. This detailed report is followed by a statement, published in 1984, by the most distinguished senior physicist working at Frascati in competition with Zichichi, Marcello Conversi, whose scientific activity started in the late forties with the discovery (together with Pancini and Piccioni) that the muon was not the Yukawa particle: an episode recalled by T.D. Lee — in his contribution to this volume — as being the first step towards the physics of leptons.

An interesting contribution is given by André Petermann with his analysis of two reports: both presented at the International Conference on the "History of the Original Ideas and Basic Discoveries in Particle Physics". One by A. Zichichi, who thought of the existence of the HL, identified its best signature, its best production process, invented and developed the needed technologies to prove its existence, and the other by M. Perl who discovered it.

There are then the comments by two great experts in the field of heavy lepton searches with (e^+e^-) colliders, Björn Wiik and Günter Wolf.

The volume closes with two Appendices. Appendix A is the reproduction of the report by A. Zichichi [presented at the International Conference on "The History of Original Ideas and Basic Discoveries in Particle Physics" 1994, H.B. Newman and T. Ypsilantis Eds., Plenum Press (1996), Vol. 352, 227-273] where it is illustrated how he developed his idea of a new Heavy Lepton (HL) and the technologies needed in order to make the search for his HL experimentally feasible. Appendix B covers the problem of priorities and is a synthesis of his review paper ["Ten Years of Work for the Third Lepton", prepared in view of the Celebrations of the Centenary of the Italian Physical Society (SIF)] in order to provide the reader with the correct answers to claims for HL and criticisms of the BCF set-up, corroborated by the references to the original papers.

The origin of the Third Family of fundamental particles is probably the best example of the lesson bestowed on us by two great leaders of our time: P.M.S. Blackett and I.I. Rabi.

Here are two of their statements.

P.M.S. Blackett

"We experimentalists are not like theorists: the originality of an idea is not for being printed in a paper, but for being shown in the implementation of an original experiment."

I.I. Rabi

"Physics needs new ideas. But to have a new idea is a very difficult task: it does not mean to write a few lines in a paper. If you want to be the father of a new idea, you should fully devote your intellectual energy to understand all details and to work out the best way in order to put the new idea under experimental test.

This can take years of work. You should not give up. If you believe that your new idea is a good one, you should work hard and never be afraid to reach the point where a new-comer can, with little effort, find the result you have been working, for so many years, to get.

The new-comer can never take away from you the privilege of having been the first to open a new field with your intelligence, imagination and hard work. Do not be afraid to encourage others to pursue your dream. If it becomes real, the community will never forget that you have been the first to open the field."

A. Zichichi started his scientific activity with P.M.S. Blackett who wanted this young fellow in his cosmic ray group; I.I. Rabi has been a strong supporter of Nino's projects and activities. The Third Family has its roots at CERN, where the insight of these two great leaders has been best implemented by A. Zichichi through his more than ten years of dedicated work to the problem of searching for a heavy lepton carrying its own leptonic number and being coupled to its own neutrino. Without the engagements of P.M.S. Blackett and I.I. Rabi, CERN would not have existed, in the same way that Modern Physics in Italy would not have flourished without Enrico Fermi. This volume is dedicated to these founding fathers of European Physics.

Tsung Dao Lee

HEAVY LEPTONS

Department of Physics, Columbia University, New York, USA

HEAVY LEPTONS

Tsung Dao Lee

Department of Physics, Columbia University, New York, USA

The history of heavy leptons is closely related to the scientific research of A. Zichichi. In this paper, I will begin with a brief history of the first heavy lepton, the muon, and then describe Nino's contribution, leading to the discovery of the second heavy lepton.

1 — Discoveries from Cosmic Radiation.

The first conclusive evidence of the existence of a particle of mass intermediate between the proton and the electron was obtained by J.C. Street and E.C. Stevenson in 1937 from their cosmic ray research [*Physical Review, 52, 1003 (L) (1937)*]. The mass value given by Street and Stevenson for the new cosmic ray particle was approximately 130 times the rest mass of the electron, which was not accurate. A much better determination (of about "240 electron masses") was given by S.H. Neddermeyer and C.D. Anderson the next year [*Physical Review, 54, 88 (1938)*].

At that time these discoveries gave substantial support to the meson theory of strong nuclear forces proposed by H. Yukawa in 1935. Moreover when what we now call the "muon" was discovered, it was thought by most physicists to be the particle theoretically predicted by Yukawa. Because of World War II, research on pure physics was interrupted. Soon after the end of the war, an important experiment was conducted by three young Italian physicists.

2 — First Identity Crisis (π, μ).

In 1947, Conversi, Pancini and Piccioni stopped the negative meson in carbon [*Physical Review, 71, 209 (L) (1947)*]. They found that the decay probability is about equal to the capture probability by nuclei. This result led immediately to the first identity crisis. In describing this crisis, we shall also identify the origin of the terms "muon" and "pion".

Soon after, Fermi, Teller and Weisskopf made a theoretical analysis. Their paper, entitled "The Decay of Negative Mesotrons in Matter", was the first in which the symbol μ was introduced. (It stood for the mesotron.) However, as we can see from the following equations (1) and (2), in today's notation the role that μ played would be replaced by the pion, the meson proposed by Yukawa. Indeed, Fermi, Teller and Weisskopf found that there is a factor of about 10^{12} discrepancy between theory and experiment.

THE DECAY OF NEGATIVE MESOTRONS
IN MATTER

E. FERMI, E. TELLER, *University of Chicago, Chicago, Illinois*
and V. WEISSKOPF, *Massachusetts Institute of Technology, Cambridge, Massachusetts*
(Received February 7, 1947)
«Phys. Rev.» 71, 314-315, (1947).

In a recent experiment Conversi, Pancini, and Piccioni [1] observed separately the behavior of positive and negative mesotrons coming to rest in iron or in graphite.

According to the conventional mesotron theories, one will have to assume that the capture now proceeds according to one of the following schemes:

(1) $$P + \mu^- = N + h\nu$$

(2) $$X + \mu^- = N + Y.$$

Here P and N stand for proton and neutron, μ signifies the mass of the mesotron, $h\nu$ is a light quantum, and X and Y stand for initial and residual nuclei in the capture process.

The experimental result [1] leads to the conclusion that the time of capture *from the lowest orbit of carbon is not less than the time of natural decay,* that is, about 10^{-6} second. This is in disagreement with the previous estimate by a factor of about 10^{12}. Changes in the spin of the mesotron or the interaction form may reduce this disagreement to 10^{10}.

A few months later Lattes, Occhialini and Powell resolved the problem by the discovery of $\pi \rightarrow \mu$ through the use of photographic emulsion [*Nature, 160, 454 (1947)*]. Since μ had already been used for the cosmic ray mesons (or mesotrons), Lattes, Occhialini and Powell introduced the symbol π which stands for "primary". In the cartoon entitled "Oh, What A Beautiful π !", a waiter in a restaurant is serving a pie, marked "MU". Rabi, Yukawa and Fermi sit around a table.

> Yukawa: Oh, what a beautiful π !
>
> Fermi: It looks weak to me.
>
> Rabi: Who ordered that ?

3 — Universal Fermi Interaction.

With the identity crisis resolved, μ became a member of a larger family. In 1949, the Universal Fermi Interaction was discovered and the Intermediate Boson Hypothesis proposed by Lee, Rosenbluth and Yang [*Physical Review, 75, 905 (1949)*].

The intermediate boson was called W, which stands for the Weak Interaction. The same Universal Fermi Interaction was independently found by G. Puppi [*Nuovo Cimento, 6, 194 (1949)*] and J. Tiomno and J.A. Wheeler [*Reviews of Modern Physics, 21, 153 (1949)*].

The question of the e-spectrum from the μ decay came under intensive experimental investigation. The results are summarized in the following two figures. In the first, the

electron distribution is plotted against

$$x = (\text{momentum of } e) \,/\, (\text{its maximum value}) \,.$$

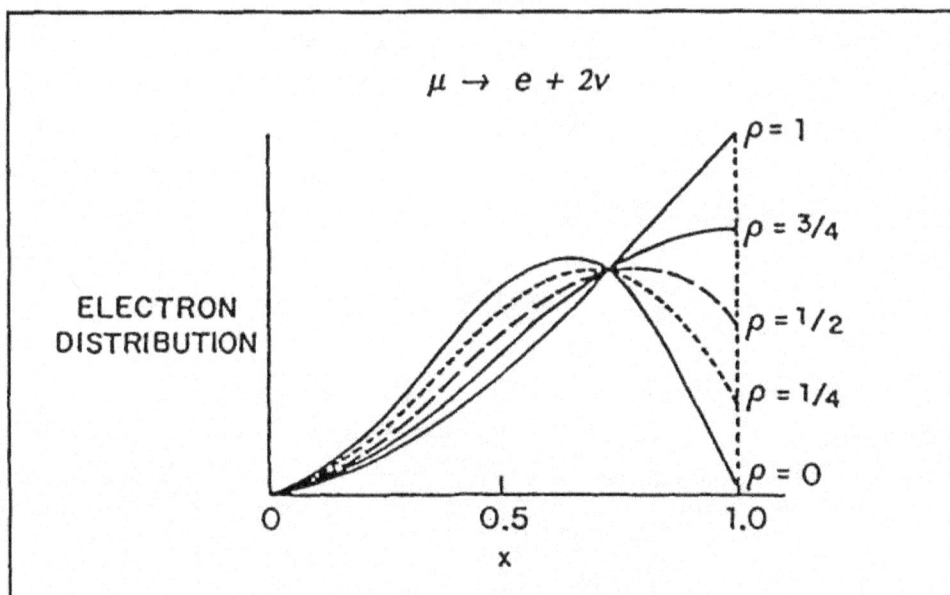

The distribution of electron energy in μ-decay.

This distribution can be characterized by the well-known Michel parameter ρ, which can be any real number between 0 and 1. The ρ value measures the height of the endpoint of the electron distribution at the maximum electron momentum $x = 1$. All of the different curves in the first figure have equal areas. As we can see, different ρ values give quite different distributions.

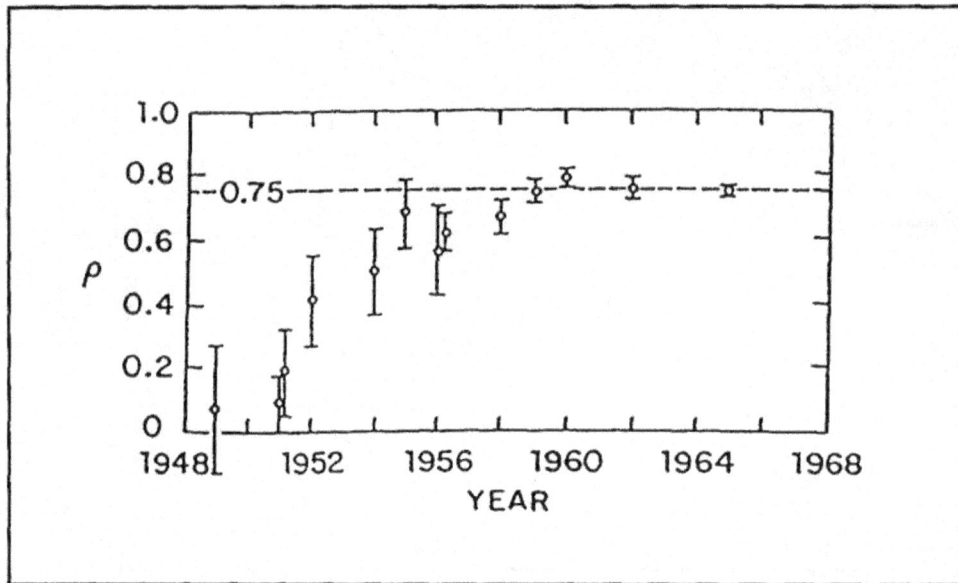

The change of the Michel parameter ρ from year to year.

In the second figure we plot the experimental value of ρ against the year when the measurement was made. The first such measurement was made in 1949 and ρ was found to be near 0. Subsequent experiments, however, yielded different values which slowly drifted upward. Only after parity non-conservation in 1957, when the two-component theory predicted $\rho = 3/4$, did the experimental value also begin to converge, finally reaching excellent agreement with theory in the sixties. What is remarkable is that, in spite of the very large difference between $\rho = 0$ and $\rho = 3/4$, as shown in this second figure, at no time did the "new" experimental value lie outside the error bars of the preceding one.

4 — Second Identity Crisis ($\theta - \tau$).

In the early 1950s, θ referred to the meson which decays into 2π, whereas τ referred to the one decaying into 3π:

$$\theta \rightarrow 2\pi$$

and

$$\tau \rightarrow 3\pi \ .$$

The spin-parity of θ is clearly $0^+, 1^-, 2^+$, etc. As early as 1953, Dalitz had already pointed out that the spin-parity of τ can be analyzed through his Dalitz plot and, by 1954, the then-existing data were more consistent with the assignment 0^- than 1^-. Although both mesons were known to have comparable masses (within $\sim 20\ MeV$), there was, at that time, nothing too extraordinary about this situation. The masses of θ and τ are very near three times the pion mass, the phase space available for the θ decay is much bigger than that for τ decay; therefore one expects the θ decay rate to be much faster. However, when accurate lifetime measurements were made in 1955, it turned out that θ and τ have the same lifetime (within a few percent, which was the experimental accuracy). This, together with a statistically much more significant Dalitz plot of τ decay, presented a very puzzling picture indeed. The spin-parity of τ was determined to be 0^-; therefore it appeared to be definitely a different particle from θ. Yet, these two particles seemed to have the same lifetime, and also the same mass. This was the $\theta - \tau$ puzzle.

The first identity crisis was caused by mis-identifying the cosmic ray mesotron as the meson hypothesized by Yukawa. It was resolved by the experimental discovery of two distinct particles μ and π. The second identity crisis was created by the misconception that θ and τ had to be two different particles. It was resolved by the breakthrough of parity nonconservation, which revealed that these two were in reality simply the different decay modes of the same particle, the kaon.

5 — Parity Nonconservation.

With parity nonconservation and the introduction of the two-component theory [T.D. Lee and C.N. Yang, *Physical Review*, *104*, 254 *(1956)*, *105*, 1671 *(1957)*], the ρ value was determined to be 3/4.

6 — The Idea That Led to the Second Heavy Lepton.

With the discovery of the two neutrinos in 1962, the physics community firmly believed that the leptons consisted of two families $(e,\ v_e)$ and $(\mu,\ v_\mu)$, each carrying its own leptonic number. While the idea of excited electrons e^* and excited muons μ^* had been discussed after the early work on ep scattering by Hofstadter, these excited leptons did not carry additional quantum numbers and therefore would decay into $e\gamma$ and $\mu\gamma$ respectively. Extensive experimental searches for these excited leptons were carried out, but none was successful.

It was A. Zichichi who had the foresight to propose the possibility of a Third Family of leptons, called the heavy lepton H_ℓ and its neutrino v_{H_ℓ}. Because of the conservation law of the new third leptonic number, these heavy leptons had to decay as follows:

$$H_\ell^+ \rightarrow \begin{cases} e^+ + \nu_e + \bar{\nu}_{H_\ell} \\ \\ \mu^+ + \nu_\mu + \bar{\nu}_{H_\ell} \end{cases}$$

with its charge conjugate reactions for H_ℓ^-. In the epoch-making paper [*INFN/AE-67/3, 20 March 1967*], Zichichi also suggested a new method to search for the heavy leptons, through $e^+e^- \rightarrow H_\ell^+ H_\ell^- \rightarrow e^+\mu^-$ or $e^-\mu^+$ (in which the two neutrinos in each of the H_ℓ decays naturally escape detection).

In addition to the bold theoretical idea of a third leptonic number, Zichichi's experimental proposal is particularly ingenious because the apparent reaction $ee \rightarrow e\mu$ seems to violate lepton conservation which, in a 4π detector environment, would drastically reduce the background. The observation of $e\mu$ coincidence, which exhibits a threshold behavior, can serve as an unmistakable signal of H_ℓ.

This proposal had an immediate impact on many laboratories. Zichichi himself conducted a systematic search for heavy leptons throughout the sixties and seventies at Frascati. These pioneering experiments showed clearly the feasibility of his proposal that the background to the apparent reaction $ee \rightarrow e\mu$ could be reduced to a negligible level [*Lettere al Nuovo Cimento, 4, 1156 (1970), 17A, 383 (1973); La Rivista del Nuovo Cimento, 4, 498 (1974)*]. However, because of the energy limitation of the ADONE accelerator, the search only led to a lower mass limit of the heavy lepton.

With the completion of the SPEAR accelerator at Stanford, which had higher energy, Martin L. Perl was able to utilize the same method on the data collected from the MARK I collaboration. He found evidence for the production of $e\mu$ coincidence events, which led to the discovery of the heavy lepton, now called τ [*Physical Review Letters, 35, 1489 (1975)*].

The discovery of τ confirmed Nino's heavy lepton idea. In addition, the fact that leptons consist of three families enforces the further theoretical speculation that the quarks should also be comprised of three families. Lederman and collaborators at FNAL soon found evidence for the charge $-1/3$-quark (b) belonging to the Third Family, and the search for the t-quark has recently produced the first evidence.

Nicola Cabibbo

LEPTON PHYSICS AT CERN AND FRASCATI

From

Lepton Physics at CERN and Frascati - Edited by Nicola Cabibbo
World Scientific Series in 20th Century Physics - Vol. 8 - 1994 - pp. xiii-xiv

WORLD SCIENTIFIC
SINGAPORE - NEW JERSEY - LONDON - HONG KONG
1995

LEPTON PHYSICS AT CERN AND FRASCATI

Nicola Cabibbo

I first met Nino when I was a student and he had an INFN fellowship. A few years later he joined the cosmic-ray group of Blackett and then he went to work at the CERN SC for the muon (g−2) experiment, which provided the first quantitative test of the leptonic nature of the muon. Nino's interest to understand the nature of the known leptons brought him to extend his experimental investigation into the new CERN machine, the PS.

It was on this occasion that I had the pleasure of the first scientific collaboration with Nino on a theoretical paper, together with Sam Berman and Raoul Gatto, to evaluate the possibility of using the time-like photons produced in proton-antiproton annihilation. This paper was the basis for the PAPEP (Proton-AntiProton Electron Pair) experiment to measure $p\bar{p}$ annihilation into electron-positron pairs.

The physics of that time was "hadron-dominated" and to propose a study of lepton pairs produced in hadronic interactions was out of the mainstream. Nino had found however a friendly audience in Victor Weisskopf, the new Director General of CERN. A year after, Nino extended the proposal to include muon detectors. Thus PAPEP became PAPLEP (Proton-AntiProton LEpton Pair). To study the production of time-like photons in hadronic interactions was his dream and I remember how much his colleagues were sceptical about the possibility of studying difficult final states, such as $(e^{+}e^{-})$ and $(\mu^{+}\mu^{-})$ pairs produced in hadronic processes. The addition of expensive muon detectors was not easily justified by the doubling of the detection power. What Nino had in mind was to search for $(e\mu)$ pairs, but this — at the time — far-fetched idea would not have passed the experimental Committees and had to remain unofficial. Viki Weisskopf, of course, knew the real reason for the "gigantic" set-up built for the simultaneous detection of electrons and muons, and with his help Nino succeeded in convincing the CERN Committees to finance the PAPLEP experiment. The idea of a charged Heavy Lepton, HL, carrying a new lepton number and accompanied by its own neutrino, ν_{HL}, had no theoretical motivation. Nevertheless it was Nino's idea and his first priority in his experimental work at CERN. He was thinking — while still working at SC — on how to detect $(e\mu)$ pairs in hadronic processes and the most difficult step was the electron detector. The π/e rejection needed an improvement of at least an order of magnitude. In 1963 Nino invented what is now called the "preshower" technique, currently used in all experiments where electrons are to be identified. The difficulties with the muon detector were its large size and the series of accurate range measurements, plus π/μ rejection studies, needed. The PAPLEP set-up was the first large solid-angle detector ever built in the world: impressive when compared to the standards of that time. Many were convinced that its dimensions were so huge and its electron and muon identification systems so complex that it would never have worked. Contrary to the predictions of Nino's detractors, his powerful detector performed perfectly well already in 1964 and a year later he published a paper where the possibility of simultaneous investigation of (e^+e^-) and $(\mu^+\mu^-)$ pairs, therefore $(e\mu)$, was established on firm experimental grounds. The enormous hadronic background had indeed been mastered. It took nearly five years of hard work coupled with technological developments in electron and muon detection in order to have the "gigantic" set-up working in a real experiment. Nino's idea of studying $(e\mu)$ final states to reduce the background in the search for new physics dates back to 1960 and an indirect trace of my discussions with him on the importance of the $(e\mu)$ final state is in my paper with R. Gatto (Physical Review 1961) on electron-positron physics where we discussed that final state as a test of W-pair production.

Having found that time-like photons were very rare in hadronic interactions, Nino moved to Frascati and, as from 1967, he went on carrying an intensive research programme on the Heavy Lepton at ADONE, using his method based on the detection of acoplanar $(e\mu)$ pairs. Had the ADONE beam energy been 20% higher, the Heavy Lepton would have been discovered by A. Zichichi at Frascati, thus crowning a scientific engagement that he had been pursuing for many years. This is in fact typical of Nino's qualities: when he is convinced that something should be done, he never gives up. The best way to understand the nature of the muon was for him to search for an even heavier lepton, not an "excited" electron or muon, but one carrying its own leptonic number. This original idea of Nino has been the driving force for his work both at CERN and at Frascati.

The interested reader can find the details of this work in a special volume "Lepton Physics at CERN and Frascati" [World Scientific, 20th Century Physics Series, Vol. 8 (1994)] which includes — together with a few selected statements by Gilberto Bernardini, Richard L. Garwin, Victor F. Weisskopf and Marcello Conversi — the reprints of the articles and of some published laboratory reports, documenting the extraordinary constancy of Antonino Zichichi in a scientific programme which spanned more than a decade and laid the foundations for the subsequent discovery by M. Perl of the Heavy Lepton HL (now called τ), first proposed and searched for at CERN and Frascati by Nino in the early sixties.

Victor F. Weisskopf

SEARCH FOR HEAVY LEPTONS
FROM TIME-LIKE PHOTONS AT CERN

From

Lepton Physics at CERN and Frascati - Edited by Nicola Cabibbo
World Scientific Series in 20th Century Physics - Vol. 8 - 1994 - pp. 45-47

WORLD SCIENTIFIC
SINGAPORE - NEW JERSEY - LONDON - HONG KONG
1995

SEARCH FOR HEAVY LEPTONS
FROM TIME-LIKE PHOTONS AT CERN

Victor F. Weisskopf

Once Nino came to my office to tell me about his ideas of studying lepton pair production at PS. I was still not Director General, but Research Director at CERN. In addition to (e^+e^-) and $(\mu^+\mu^-)$ pairs, he wanted to search for $(e^\pm\mu^\mp)$ pairs as a signature of a new lepton carrying its own lepton number. He told me that if such a lepton existed with one GeV mass, it would have escaped detection in hadron accelerator experiments for two reasons: i) it would decay with a lifetime of order 10^{-11} sec and ii) because there is no $\pi \to \mu$ mechanism for such a heavy new lepton: for its production a time-like photon would be needed. Time-like photons could be produced in hadronic interactions: for example in $(\bar{p}p)$ annihilation. This was before Lederman-Schwartz and Steinberger had discovered the two neutrinos. To think of a "sequential" Heavy Lepton and to work out the possible ways to get it in a hadron machine was for me extremely interesting. Nino had just finished his first high precision work on the muon (g–2). It was some time after the Rochester Conference in 1960. I gave Nino the following suggestion: if you want to search for something so revolutionary as a Heavy Lepton carrying its own lepton number you should work out a proposal for a series of experiments where the study of lepton pairs (e^+e^-) and $(\mu^+\mu^-)$ could be justified in terms of physics accepted by the community. In addition a high intensity antiproton beam was needed. He came later to tell me that he had two very good friends, both excellent engineers: Mario Morpurgo and Guido Petrucci. A very high intensity antiproton beam could be built to study the electromagnetic form factor of the proton in the time-like region. If the proton was "point-like" in the time-like region, the rate of time-like photons yielding (e^+e^-) and $(\mu^+\mu^-)$ pairs could be accessible to experimental observation, thus allowing to establish some limits on the new Heavy Lepton mass, or to see it, via the $(e^\pm\mu^\mp)$ channel.

The "official" theme was: to establish if the proton had a structure or not in the time-like region. Thus a powerful system able to detect (e^+e^-) and $(\mu^+\mu^-)$ pairs could be built. Nino established in 1963 the existence of a time-like structure of the proton studying the (e^+e^-) channel and in 1964 studying the $(\mu^+\mu^-)$ channel. The set up was able to do what he wanted: a simultaneous detection of electrons and μ pairs, therefore $(e^\pm\mu^\mp)$ as well. Unfortunately the proton was not a point-like particle in the time-like region and therefore the source of time-like photons originated in $(\bar{p}p)$ annihilation was very depressed. In fact, using the (e^+e^-) and the $(\mu^+\mu^-)$ channels, Nino established that at 6.8 $(GeV/c)^2$ time-like four momentum transfer, the cross-section was 500 times below the expected point-like value. This result had attracted a lot of attention. Bogoliubov was very interested when in 1964 Nino went to Dubna to present the $(\mu^+\mu^-)$ results at the International Conference on "High Energy Physics". Yang had a model that predicted a point-like structure of the proton in the time-like region. I called this series of experiments as measuring the "heartbeat of the proton". Of course there were no $(e^{\pm}\mu^{\mp})$ events, neither in the $(\bar{p}p)$ nor in the (π^-p) channel. Nevertheless a series of experiments was performed on "standard" physics, such as the discovery of many rare decay modes of mesons and the measurement of the $(\omega-\phi)$ mixing.

All these experiments could be done because Nino had invented what is now known as the "preshower" method to reject with high efficiency pions in favor of "electrons". Once it was clear that in hadronic interactions there are very few time-like photons, he asked me if I would give the green light in order to consider the use of the $(e^{\pm}\mu^{\mp})$ technology in the newly being developed Frascati (e^+e^-) collider. There the "time-like" photons were very abundant and the $(e^{\pm}\mu^{\mp})$ method would have been the best in order to see if a Heavy Lepton carrying its own lepton number existed. Of course he got the green light and in 1970 he got the first limit on the Heavy Lepton mass together with a series of high precision QED measurements.

Samuel C.C. Ting

THE BEGINNING OF THE PHYSICS OF LEPTONS

Massachusetts Institute of Technology, Cambridge, MA, USA

THE BEGINNING OF THE PHYSICS OF LEPTONS

Samuel C.C. Ting

Massachusetts Institute of Technology, Cambridge, MA, USA

ABSTRACT

Over the last 30 years the study of lepton pairs from both hadron and electron accelerators and colliders has led to the discovery of J, Υ, Z and W particles. The study of acoplanar $e\mu$ pairs + missing energy has led to the discovery of the heavy lepton, now called τ lepton. Indeed, the study of lepton pairs with and without missing energy has become the main method in high energy colliders for searching new particles.

This paper presents some of the important contributions made by Antonino Zichichi over a 10 year period at CERN and Frascati in opening this new field of physics. This includes the development of instrumentation to distinguish leptons from hadrons, the first experiment on lepton pair production from hadron machines, the precision tests of electrodynamics at very small distances, the production of hadrons from e^+e^- collisions and most importantly his invention of a new method $e^+e^- \rightarrow e\mu +$ missing momenta, experimentally proving that, thanks to his new electron and muon detection technology, these signals have very little background.

THE BEGINNING OF THE PHYSICS OF LEPTONS

Samuel C.C. Ting

The study of lepton pair production from hadron collisions was originated by A. Zichichi. Indeed, the first publication on this subject appeared in *Physics Letters, 5, 195 (1963)*. At that time, the physics community was mostly interested in studying strong interactions such as the Chew-Low plots and Regge trajectories. The physics program in the United States, at Brookhaven, Argonne and Berkeley as well as at CERN, was devoted to the study of strong interactions and neutrino physics.

To carry out lepton pair study, a revolutionary detector had to be made as shown in Fig. 1 below.

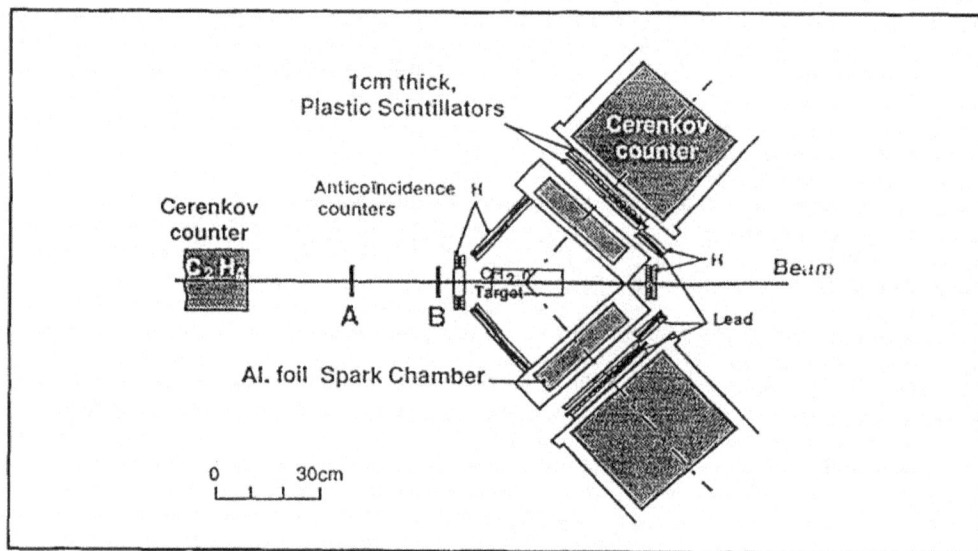

Fig. 1: Experimental layout for the first experiment on $\bar{p} + p \rightarrow e^+ + e^-$.

This detector covered very large solid angle in the center mass system, but most importantly, had a very strong rejection power against pions. Zichichi and his collaborators had to develop pre-shower methods [*CERN 63-25, Nuclear Physics Division, June 27, 1963*] to obtain a e/π rejection of 10^{-3}. The spectrum they obtained by the pre-shower technique is shown in Fig. 2. The importance of this development comes from the fact that in the late 1950's and early 1960's there were only two instruments to distinguish electrons and pions in the GeV region: the lead glass total absorption Cerenkov counter [G. Gatti et al., *Review of Scientific Instruments, 32, 949 (1961)*] and the total absorption scintillation detector [G. Backenstoss et al., *NIM, 20, 294 (1963)*]. Both instruments have a rejection power e/π of a few per cents. The pre-shower technique of Zichichi is achieved by looking at the first stage of the shower development as it occurs in a lead slice of two radiation lengths.

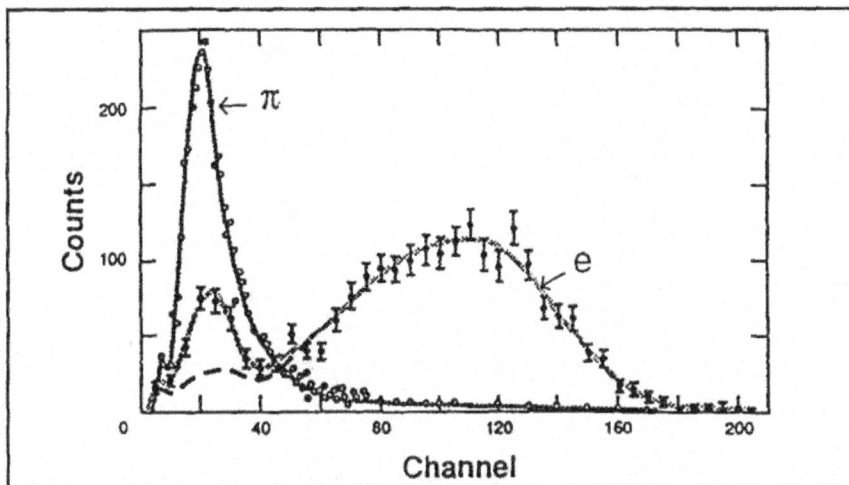

Fig. 2: Pulse height spectra with pre-shower technique. This allows one to reach an e/π rejection of 10^{-3}.

The Zichichi group also pioneered the study of muon "punch through" [*CERN 64-31, June 24, 1964* and *Nuovo Cimento, 35, 759 (1965)*] in order to improve muon detection technology. Continuing the development of instrumentation, Zichichi and his collaborators found a way to improve the e/π rejection to order of 10^{-4} (Figs. 3 and 4) [see *Nuovo Cimento, 39, 464 (1965)*].

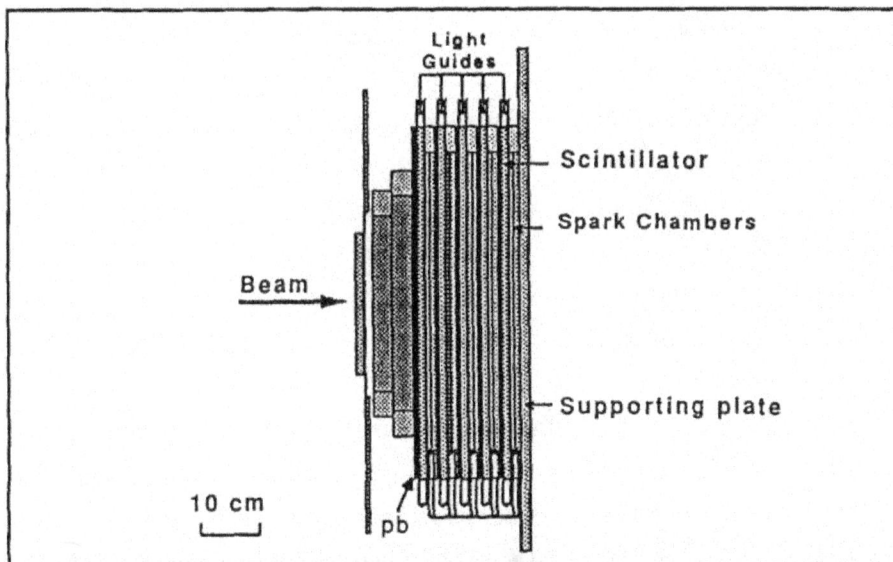

Fig. 3: Electron detector, which consists of five elements, each one being made of a lead layer followed by a plastic scintillation counter and a two-gap spark chamber.

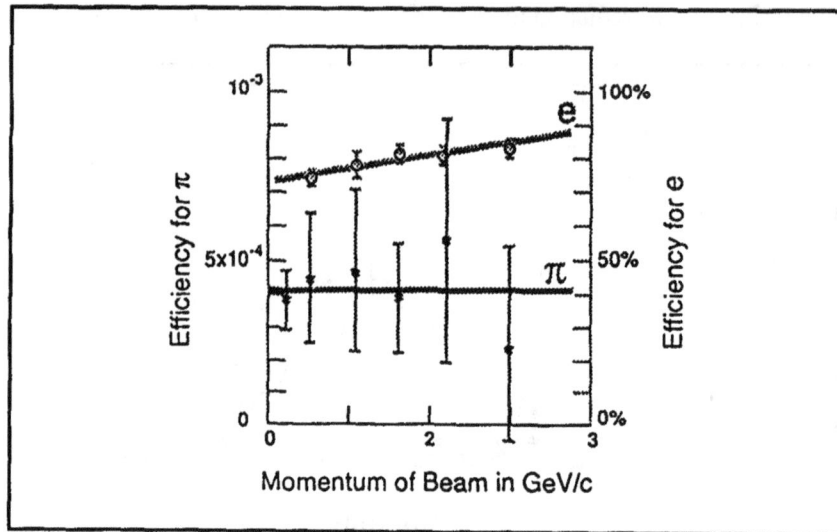

Fig. 4: The rejection power of this new detector against pions is of the order of a 4×10^{-4} (left scale) up to 2.5 GeV. The efficiency for electron detection varies from 75% to 85% (right scale).

The simultaneous detection of electrons and muons was the basic feature of the PAPLEP (Proton-AntiProton annihilation into LEpton Pairs) set-up at CERN, where leptonic final states were investigated with a very large solid angle detector and high rejection power distinguishing electrons and muons from hadrons. This detector is shown in Fig. 5 hereafter and was originally presented by Zichichi at the Dubna Conference in 1964 [*Atomizdat, Moscow, 1966, Vol. 1, 857*] and published in 1965 [*Nuovo Cimento, 40, 690 (1965)*].

Let me quote from the above reference the following statement (pages 691-693): "The electronic signal triggered the thin-plate spark chambers (*for track detection*), the electron detectors (*for electron detection*) and the range chambers (*for muon detection*) which were placed after the counters H". This is on line with the testimony of Weisskopf and Cabibbo [see V.F. Weisskopf, *The heartbeat of the proton*, in "Lepton Physics at CERN and Frascati", N. Cabibbo Ed., 20th Century Physics Series, Vol. 8, World Scientific, 1994, p. 45 and N. Cabibbo, *Foreword*, in the same volume, p. xiii] on the purpose of the PAPLEP experiment at CERN.

It is important for us to realize that the physics of lepton pair production from hadron collisions and the development of instrumentation have opened an entire new field which eventually led to the discoveries of the *J* particle, Upsilon particle, *Z* and *W*.

The PAPLEP experiment established the validity of the new technologies for electron and muon detection to master the background, while pointing out that $(\bar{p}p)$ was a very poor source of time-like photons. The Zichichi group abandoned the muon channel and concentrated on the electron channel to carry out a series of experiments on meson physics with (e^+e^-)

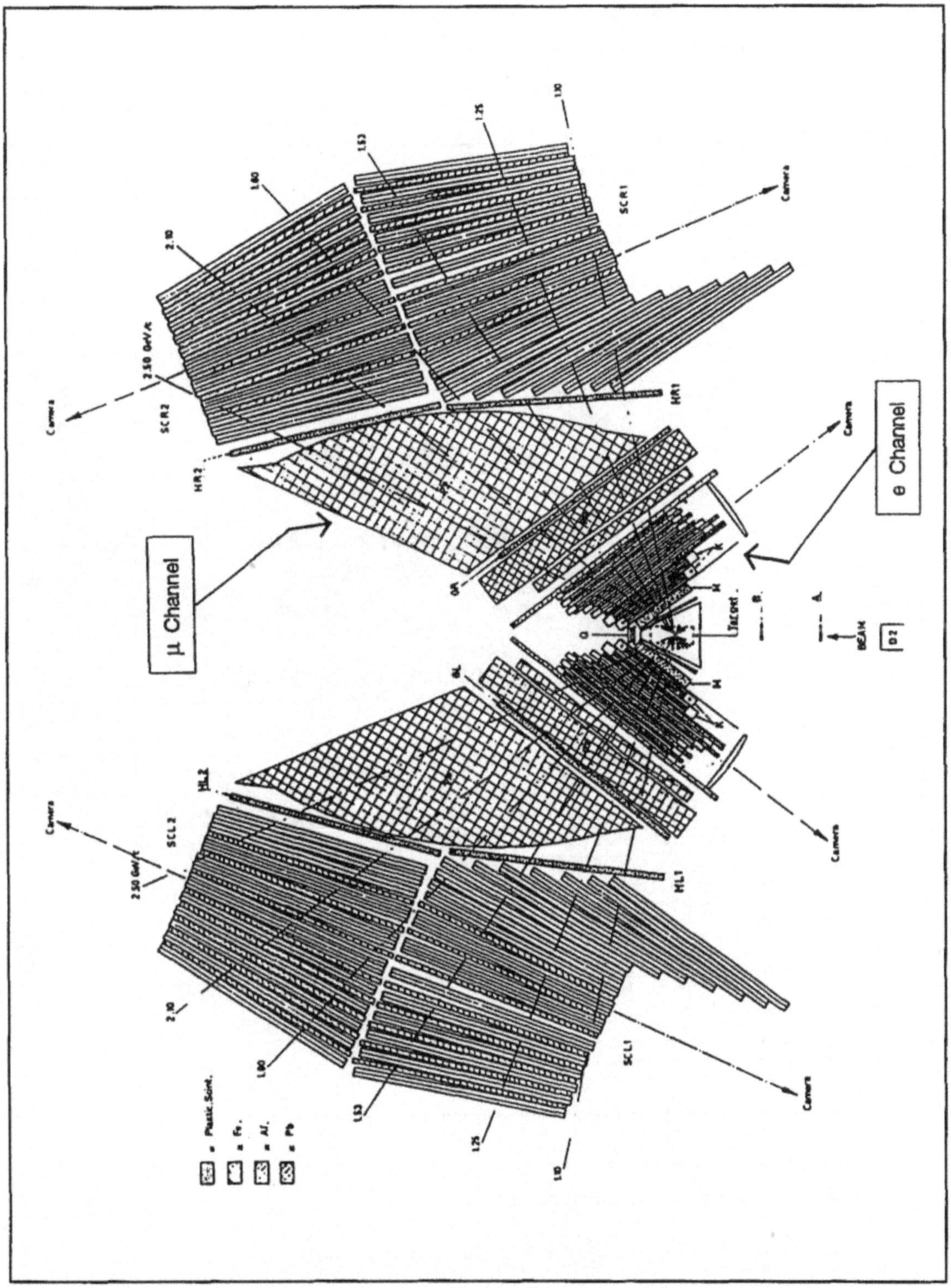

Fig. 5: The PAPLEP experimental set-up. For the scale, see Fig. 3 where the structure of the electron detector and its dimensions are shown.

final states and in particular the difficult experiment of measuring the ω–φ mixing angle [see *Nuovo Cimento, 57A, 404, (1968)*] with a detector shown in Fig. 6 and results shown in Fig. 7. This established the generalized Weinberg Sum Rule.

Fig. 6: Photograph of the experimental set-up, showing the scale of the detector. The neutron counters are mounted on rails.

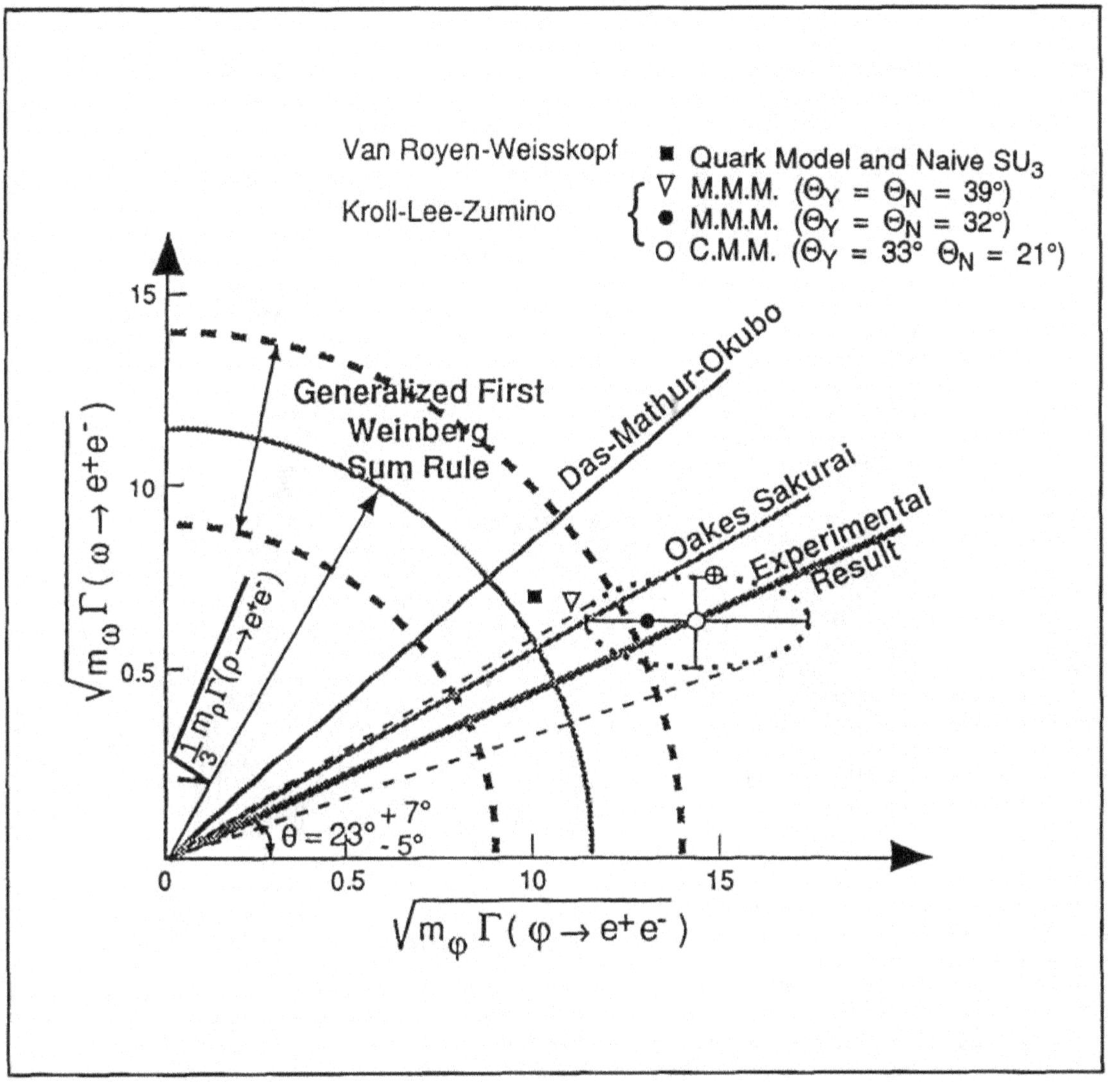

Fig. 7 : Experimental results compared with theoretical predictions.
The heavy circle indicates the relation existing between the (e^+e^-) decay widths of the three vector mesons ρ, ω, φ on the basis of the generalized first Weinberg sum rule. The dashed circles indicate the error limits due to the uncertainty in $\Gamma(\varphi \to e^+e^-)$ and m_ρ. The two points indicated with MMM correspond to the two versions of the Mass-Mixing Model of Kroll, Lee and Zumino while the CMM point is the prediction of their Current Mixing Model. The quark model is that of Van Royen and Weisskopf.

When the highest energy electron-positron collider, ADONE, was constructed in the second half of the 1960's, Zichichi began a comprehensive study of all final states from e^+e^-

collisions with a large solid angle detector (Fig. 8) to systematically study the following reactions [Proceedings of the VIII course "Ettore Majorana" International School of Subnuclear Physics, Erice 1970, published by Academic Press, New York-London, 1971]:

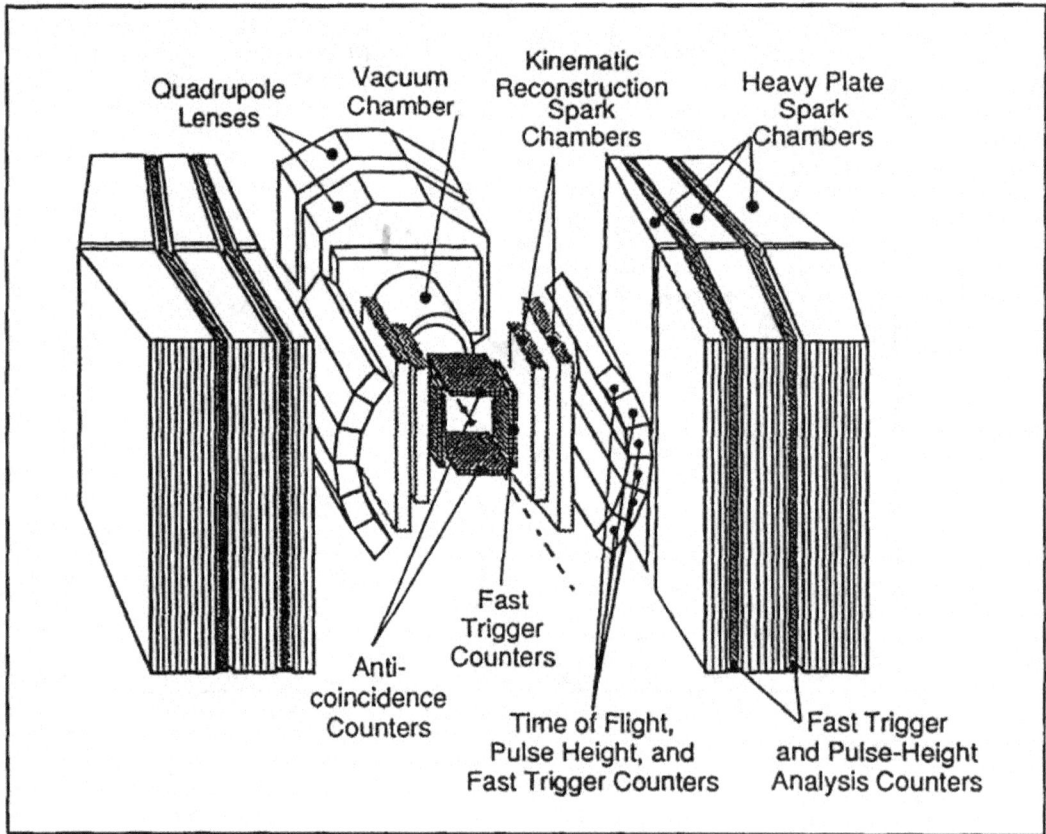

Fig. 8: Perspective of the experimental set-up.

a) Test of QED from $e^+e^- \rightarrow e^+e^-$ or $\mu^+\mu^-$.

At that time, there was no experimental test of QED at very small distances. Without the validity of QED it would be impossible to calculate the background to search for heavy leptons. Zichichi's accurate study of the reaction: $e^+e^- \rightarrow e^{\pm}e^{\mp}$ (see Fig. 9) established the validity of QED at very small distances and gave us confidence that QED is correct including the acoplanar radiative corrections (see Fig. 10). Before the work of Zichichi et al. at Frascati, radiative corrections were computed using the so-called "peaking approximation". This produces coplanar (e^+e^-) and ($\mu^+\mu^-$) events in the final state. The acoplanarity of final states (e^+e^-) and ($\mu^+\mu^-$) produced by Standard QED [see *Physics Letters, 36B, 149, (1971)* and *Physics Letters, 45, 169, (1973)*] was of great importance because the search for heavy sequential leptons from $e^+e^- \rightarrow e\mu + X$ was based on the detection of acoplanar ($e^{\pm}\mu^{\mp}$) events.

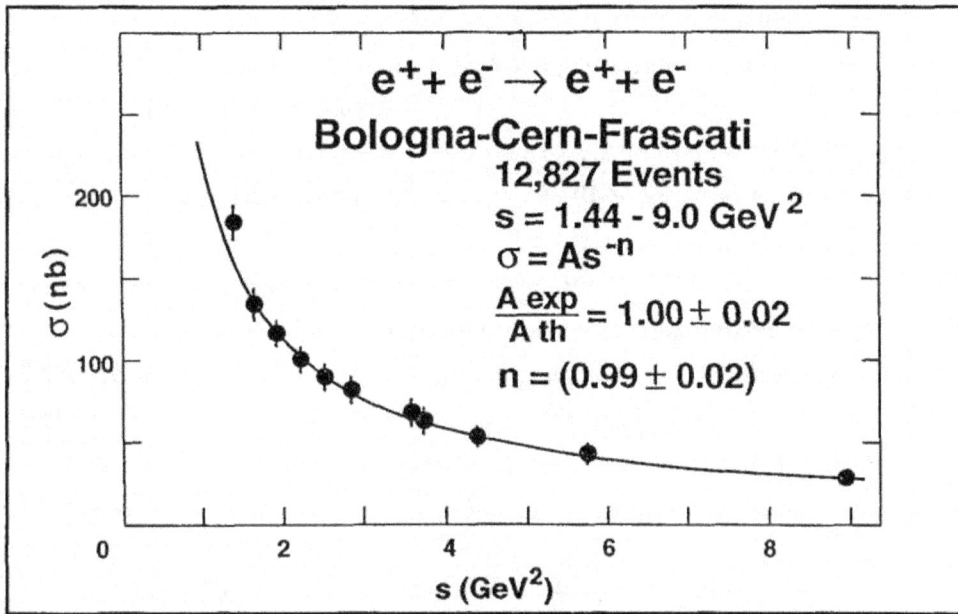

Fig. 9 :   The cross section $\sigma(e^+e^- \to e^+e^-)$ vs. s, as measured at ADONE.

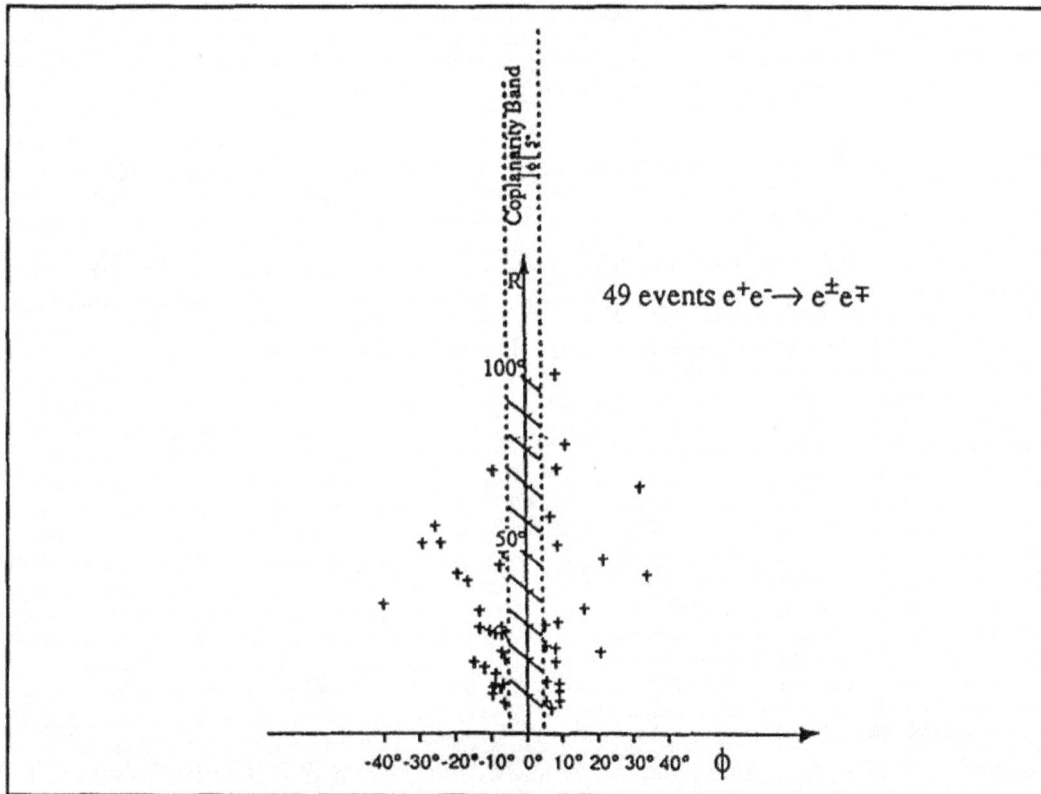

Fig. 10 :   R (the acollinearity angle) vs ϕ (the acoplanarity angle) scatter diagram for all $(e^{\pm}e^{\mp})$ events with $|\phi| > 5°$.

b) Test of lepton number conservation from $e^+e^- \to e^{\pm}\mu^{\mp}$.

This exclusive reaction also provides information on background from beam-gas interaction for the process: $e^+e^- \to e\mu$ + missing energy. In the 1970 paper by Zichichi [Proceedings of the VIII course "Ettore Majorana" International School of Subnuclear Physics, Erice 1970, published by Academic Press, New York-London, 1971], there is an important remark *"the study of this reaction allows one to establish the validity of the leptonic number selection rules for high space-like and time-like q^2 values. Collinear events with an electron and a muon in the final state would represent a proof that the presently known leptonic selection rules are violated. Background sources for this reaction are proved to be absent, from beam-gas interaction, cosmic radiation, or from simulation by $(e^{\pm}e^{\mp})$ or $(\mu^{\pm}\mu^{\mp})$ final states. No events of the type (b) were found, and from the total number of observed lepton pairs, we get:*

$$\frac{e^+e^- \to e^{\pm}\mu^{\mp}}{e^+e^- \to \text{lepton} + \text{antilepton}} \leqslant 2 \cdot 10^{-3} \ \text{with 95\% confidence"}.$$

c) $e^+e^- \to h^+h^-$ + anything.

This is the first study on hadron production from e^+e^- colliders away from low energy resonance region, which shows that hadrons are indeed produced and therefore must be rejected in a reaction: $e^+e^- \to e\mu$ + missing energy.

d) Search for sequential heavy leptons, from the reaction: $e^+e^- \to HL^+HL^- \to e^{\mp}\mu^{\pm}$ + missing energy.

When first presented in 1967 [*INFN/AE - 67/3, 20 March 1967*], this idea encountered universal criticism in the physics community. Not only most of the physicists believed heavy leptons should come in the form

$$e^* \to e\gamma,$$

but also no one believed an experiment of

$$e^+e^- \to HL^+HL^- \to e^{\mp}\mu^{\pm} + \text{missing energy},$$

could technically be carried out successfully.

Indeed most of the physicists (me included) had worked on $ep \to ep$ and $\mu p \to \mu p$ to search for e^* and μ^*.

The comprehensive program of a) b) c) d) above was based on the unique development of instrumentation, the experience in lepton pair physics and particularly in e-hadron rejection and μ-hadron rejection performed at CERN with PAPLEP. This enabled the Zichichi group to build in Frascati a detector specially designed for the items a) b) c) d), all studied as a coherent program. The results on heavy lepton search were first published in *Nuovo Cimento*, 4, 1156 (1970).

This publication stimulated many interests. It was the first publication on the following:

1. The idea of the possible existence of a heavy lepton with its own quantum number and its own neutrino.

2. The idea of searching for this type of new heavy lepton by the acoplanar $e\mu$ coincidence method from the reaction $e^+e^- \rightarrow HL^+HL^- \rightarrow e^{\mp}\mu^{\pm}$ + missing energy.

3. Most importantly, it experimentally demonstrated that the idea and its instrumentation worked cleanly.

There were many discussions in 1970 and 1971, some published in Physics Today and Physics Review, on the importance of this work. It should be noted that:

i) Because of the studies of a) and b) above and because these types of new heavy leptons were point-like particles, their productions and decays could be reliably calculated using quantum electro-dynamics and standard weak interactions phenomenology.

ii) Because of the systematic studies of a) b) and c), the backgrounds could be estimated and found to be negligible.

Zichichi gave many seminars on his original idea in Rome, at Frascati and CERN and requested more machine time to run at higher energies and higher intensities. The final results were published in *Nuovo Cimento, 17A, 383 (1973)* which I include in Table 1 and in Fig. 11.

Limit on the Mass of Heavy Leptons

Beam Energy MeV	Integrated Luminosity $(\times 10^{32}$ cm$^{-2})$
600	50
650	80
700	74
750	175
800	102
850	130
950	630
970	235
1050	1861
1200	449
1500	800

Table 1

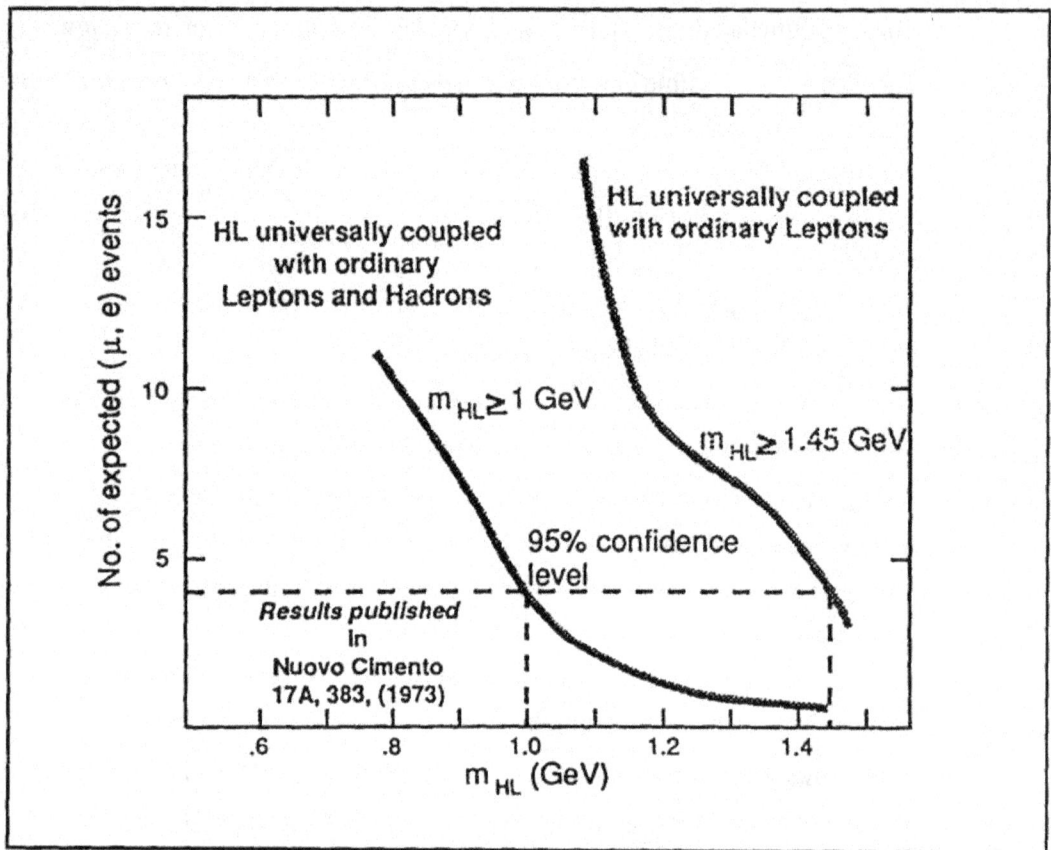

Fig. 11: Expected number of $(\mu^{\pm}e^{\mp})$ pairs vs. m_{HL} for two types of universal weak couplings of the heavy leptons.

Zichichi's development of his innovative instrumentation for electron and muon identification, his wide interest and detailed studies in lepton physics have contributed fundamentally to open this new field which turned out to be essential for basic discoveries.

Claudio Villi

THE BASIC STEPS WHICH LED TO THE DISCOVERY
OF THE HEAVY LEPTON τ: A HISTORICAL RECORD

From

Il Nuovo Cimento - Vol. 107 A, N. 5 - Maggio 1994 - pp. 665-674

EDITRICE
COMPOSITORI
BOLOGNA
1994

THE BASIC STEPS WHICH LED TO THE DISCOVERY
OF THE HEAVY LEPTON τ: A HISTORICAL RECORD

Claudio Villi

Dipartimento di Fisica dell'Università - Padova, Italia
INFN, Sezione di Padova - Padova, Italia

(ricevuto e approvato il 28 Marzo 1994)

SUMMARY

A review of the events which led to the discovery of the τ-lepton is presented. This work is based on a search of the literature and from the position of responsibility the Author had in the Institute for Nuclear Physics (INFN), responsible for the research activity with the new (e^+e^-) collider ADONE. The review follows the sequence of events leading to the discovery of the τ-lepton: from the original idea of a Heavy Lepton carrying its own leptonic number, to the invention of the acoplanar $(e\mu)$ method and associated technologies (preshower detector), to the evidence that the $(e\mu)$ method works, contrary to the prevailing opinion that the high background level could not be controlled. The review is aimed at establishing the sequence of fundamental developments which made this fascinating field possible.

1 — The origin of the idea of a Heavy Lepton with its own neutrino and the invention of the acoplanar $(e\mu)$ method.

Thinking about the discovery of the τ-lepton, and about its deep significance, I have recalled the years when I had been a witness to what was going to become one of the greatest scientific enterprises of this century. During the years 1966-1975, I was at first member of the Executive Committee, then Vice-President and finally President of INFN, the Italian State Institution responsible for research in nuclear and particle physics. During the same period, I was responsible for the research activity of the Laboratori Nazionali di Frascati, where the new project, called ADONE, was designed and constructed: at that time it was the largest $(e^+ e^-)$ collider in the world. In this capacity I was therefore closely involved with the scientific and organizational work being planned for experiments in ADONE. We were in great need of new ideas to be added to the standard experimental programme of ADONE: a programme which was based on the high-energy and high-precision tests of quantum electrodynamics and on the production of the known $(\rho\omega\phi)$ vector mesons. ADONE appeared to be an apparatus doomed to verifying things already expected. But there was then a young physicist with new ideas: Nino Zichichi. We were linked by some common physics interests. My research work with Clementel on the proton form factor had attracted his attention. In 1966, Nino Zichichi produced, together with T. Massam, a description of the electromagnetic form factor of the nucleon where he put all the then-known vector mesons and a further q^2 dependence in the (vector-meson-nucleon) vertex function. These were topics extremely interesting to me, but, when he came to Padua for a Seminar on the electromagnetic form factor of the nucleon, it was obvious that he was very eager to talk about his new ideas for physics with ADONE. I still remember one evening when we were having dinner in a «trattoria» in Rome. He brought me his proposal [1] to search for heavy leptons. He wanted to explain to me the new method that would enable him to discover whether, in addition to the electron and the muon, there was a new heavier lepton with its own neutrino. At that time, when talking about leptons, proposals were always about excited electrons, e^*, or about excited muons, μ^*, not about leptons like those that today are called «sequential».

A remark is in order: since the pioneering work on (ep) scattering by Hofstadter, many experiments on (ep) and (μp) scattering had been performed. The results were interpreted as ways to search for excited electrons (e^*) and excited muons (μ^*). These heavy leptons were in fact expected to decay into $(e\gamma)$ and $(\mu\gamma)$ respectively. In the late 60's and early 70's, many storage rings were built to continue the search for this type of heavy leptons, i.e. the e^* and μ^*. There were no ideas to search for a new type of lepton carrying its own leptonic number and therefore decaying into three bodies, thus giving rise to an apparent violation of the known leptonic numbers via the $(e\mu)$ final state. There was no experimental method proposed to search for this new type of Heavy Lepton. During the dinner, Nino insisted that if such a lepton really existed, the best way to observe it was via the measurement of an $(e\mu)$ acoplanar pair without anything else in the final state.

During the entire period of the experiment, from approval to construction, to data-taking, the atmosphere in Frascati was very negative. Nino Zichichi's critics maintained that the background would be impossible to keep under control, and that in an (e^+e^-) collider there are always plenty of electrons and spurious muons (from pion decays).

Zichichi had been working at CERN on the production of (e^+e^-) and $(\mu^+\mu^-)$ pairs in strong interactions. He had also invented a new technique (known today as «preshower») that allowed the selection of an electron-positron pair with a powerful rejection factor, 10^{-6}, against the hadron background. For Zichichi the background was not a problem.

He invented the method specifically to search for the kind of new lepton he had in mind. He was thinking about these problems while working at CERN on lepton pair production in hadronic interactions (see sect. 4). In fact in a paper published together with T. Massam in 1966 (Nuovo Cimento, <u>43A</u>, 227 (1966)), he says: *«But we know that nucleons are very poor sources of timelike photons».* If a heavy lepton existed, this would not have been easily produced in hadron collisions. In a lecture at the 1970 Erice School, Zichichi says: *«... even if heavy leptons exist, they would have escaped detection because there is no mechanism of the type $\pi \to \mu$ which would allow these leptons to be produced via a strongly interacting hadron and then remain a quasi-stable particle (like the muon)»* (in ref. [6b], p. 803). The most efficient way to produce heavy leptons was (and is) (e^+e^-) reaction: a very efficient source of timelike photons.

Thus the ADONE machine was for Zichichi a unique tool for the production of the new heavy lepton he had in mind. The basic reaction being (as reported in ref. [1]):

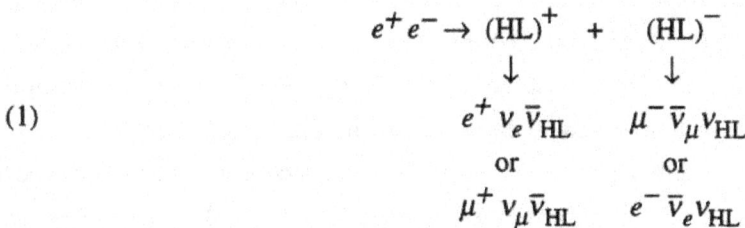

(1)

$$
\begin{array}{ccc}
e^+e^- \to & (HL)^+ & + & (HL)^- \\
& \downarrow & & \downarrow \\
& e^+ v_e \bar{v}_{HL} & & \mu^- \bar{v}_\mu v_{HL} \\
& \text{or} & & \text{or} \\
& \mu^+ v_\mu \bar{v}_{HL} & & e^- \bar{v}_e v_{HL}
\end{array}
$$

The key point was to search for acoplanar $(e^\pm \mu^\mp)$ pairs in the final state of reaction (1).

In 1970, the first results were published [2]. These results proved that the $(e\mu)$ method implemented at Frascati was working as expected and the background to reaction (1) could indeed be reduced to a negligible level.

Three years later [3] the final results supported the conclusion that the search for acoplanar $(e\mu)$ pairs produced in (e^+e^-) annihilation was the right way to follow. In fact, if a «signal» was produced, the «signal» would have been perfectly visible. All objections on the irreducible $(e\mu)$ background were proven to be unfounded by the experimental results published by A. Zichichi and his collaborators.

To summarize this first part of my recollection of events, the contributions of Zichichi are the following:

i)   proposal for the existence of a Heavy Lepton with its own neutrino,

ii) invention of the $(e\mu)$ method,

iii) proof that the $(e\mu)$ method works (contrary to other people's claims that the background would be impossible to control).

Zichichi was invited to give seminars on his original ideas, his experimental technique and his results. I was very impressed by the strong interest the theoreticians showed for his ideas and his work. I recall a presentation at Frascati during the *Informal Meeting on Recent Developments in High-Energy Physics, 26-31 March 1973* [4].

The subsequent discovery by M.L. Perl [5] of acoplanar $(e\mu)$ pairs using (e^+e^-) at energies slightly higher than ADONE, following precisely Zichichi's method, is the best proof for the validity of the new idea on the existence of a Heavy Lepton with its own neutrino, and of the importance of the invention of the acoplanar $(e\mu)$ method, implemented and shown to work at Frascati.

2 —   Further searches at higher energies promoted by A. Zichichi during the years 1972-1974.

The fact that no signal was seen at ADONE did not discourage Zichichi from going on searching at higher energies. At another memorable dinner at the same «trattoria», he presented to me all the reasons why the ADONE energy had to be increased. Not only for the Heavy Lepton search, but also for the «narrow» resonances: another topic which Nino tried to pursue vigorously at Frascati. Again the theoretical trend was negative: if a resonance exists in the GeV mass range, its width must be large, i.e. ~ 0.1 GeV. Why search for narrow resonances? In spite of the theoretical discouragement, Zichichi's interest in (e^+e^-) physics did not decline. He presented a series of reports at various conferences in Wiesbaden (1972), Batavia (1972), Pavia (1973), Frascati (1973) and Bielefeld (1973), where the many reasons why (e^+e^-) physics had to be promoted were illustrated. A review paper based on these reports was published by Zichichi in 1974 [4]; the title is self-explanatory, *Why (e^+e^-) physics is fascinating*. From this paper it appears that (e^+e^-) colliders were, according to Zichichi, a potential source of new physics, able to compete with proton synchrotrons. These were the times when the vast majority of the physics community was attracted by «hadronic» machines. In (e^+e^-) physics, only «butterflies» were expected: a trend which dominated the Frascati scientific atmosphere in the crucial years when the experimental set-ups were designed and implemented. The negative results obtained at Frascati on the search for a Heavy Lepton did not discourage Zichichi. His view was that the Frascati work was the proof that the $(e\mu)$ technique was working and that further searches had to be encouraged and implemented. In the review paper quoted above [4], Zichichi elaborated on the properties of the Heavy Lepton to be measured (decay correlations, decay spectra, decay rates) in order to ensure that the observed $(e\mu)$ pairs were indeed from Heavy Lepton decays.

3 — Basic experiments needed in order to understand the physics of acoplanar $(e^{\pm}\mu^{\mp})$ pairs.

There is a detail which is often forgotten, even if of great value in order to correctly understand Zichichi's engagement in his search for a Heavy Lepton. First of all, one had to make sure that, at high q^2 values, the lepton conservation number was valid, even though at low q^2 values ($q^2 \simeq 10^{-2}\,\text{GeV}^2$), the $(\mu \to e\gamma)$ had been checked. This is why a detailed study of the two-body reaction:

$$e^+e^- \to e^{\pm}\mu^{\mp}$$

was performed. In 1970 the limit was [6a]

$$\frac{e^+e^- \to e^{\pm}\mu^{\mp}}{e^+e^- \to \text{lepton} + \text{antilepton}} \leqslant 2 \cdot 10^{-3}$$

and the final result (1973) $\leqslant 7 \cdot 10^{-5}$ [4].

It is interesting to note that in a lecture presented at the Erice School in 1970 [6b], Zichichi, with reference to the two-body reaction $e^+e^- \to e^{\pm}\mu^{\mp}$, says: «*The study of this reaction allows one to establish the validity of the leptonic number selection rules for high spacelike and timelike q^2 values. Collinear events with an electron and a muon in the final state would represent a proof that the presently known leptonic selection rules are violated. Background sources for this reaction are proved to be absent, from beam-gas interaction, cosmic radiation, or from simulation by $(e^{\pm}e^{\mp})$ or $(\mu^{\pm}\mu^{\mp})$ final states*».

Furthermore, a careful study of standard QED processes was needed, the most important being the acoplanar radiative effects in (e^+e^-) and $(\mu^+\mu^-)$ final states produced in (e^+e^-) interactions. In fact if, at some threshold energy, the leptonic number selection rules were violated, then $(e\mu)$ coplanar pairs would be produced. The coplanarity of the $(e\mu)$ pairs could be destroyed by radiative effects, thus giving rise to acoplanar $(e\mu)$ events. Radiative effects had thus far been studied using the so-called «peaking approximation». In this approximation, QED could not produce acoplanar events in the final state. A quantitative knowledge of the acoplanar radiative effects was needed in order to study a process where acoplanar $(e^{\pm}\mu^{\mp})$ events were the main signal to detect.

Moreover, the proof was necessary that QED processes such as

$$e^{+}e^{-} \rightarrow e^{+}e^{-} ,$$
$$e^{+}e^{-} \rightarrow \mu^{+}\mu^{-} ,$$

were following the theoretical predictions, in order to be sure that what was expected was indeed correctly observed in the experimental set-up where the acoplanar $(e^{\pm}\mu^{\mp})$ events had to be observed if the Heavy Lepton was produced.

The Zichichi group is the only one who produced all this. These experimental results are of basic value in order to be sure that, if acoplanar $(e\mu)$ pairs were observed, these could not be due to QED effects coupled with leptonic number violation. It is interesting to review these papers. The first [6a] was published in 1970 and refers to the validity of the leptonic selection rules mentioned above: i.e., to the fact that QED did not permit the two-body process $e^{+}e^{-} \rightarrow e^{\pm}\mu^{\mp}$. The second [7] was published in 1971 and concerns the discovery that acoplanar radiative effects were indeed observed, thus proving the inadequacy of the «peaking» approximation in radiative corrections. The third [8] refers to the validity of crossing symmetry in the electromagnetic interactions of the known leptons. The fourth [9] is devoted to check the QED equivalence between the two known leptons: the electron and the muon. The fifth [10] is the best high-precision check of QED in $(e^{+}e^{-})$ interactions at the maximum q^{2} values available at that time. The sixth paper [11] refers to the acoplanar radiative effects measured for $(\mu^{+}\mu^{-})$ pairs produced in $(e^{+}e^{-})$ interactions. The last contribution [12] is the high-precision QED checks for the reaction $(e^{+}e^{-}) \rightarrow (\mu^{+}\mu^{-})$.

These seven papers are the best evidence that Zichichi's group was the only one really engaged in the heavy-lepton search at Frascati. This search included in fact all QED checks mentioned above. The amount of work impressed me; Zichichi kept me well informed of all progress of his group as he was constantly insisting on the great value of pushing up the ADONE energy as much as possible.

4 — Experimental work conducted by A. Zichichi during the years 1960-1968.

In order to realize the involvement of A. Zichichi in this type of physics, let me recall what he was doing in the years 1960-68, before the Frascati proposal [1] was presented.

As mentioned above, in 1965 he invented [13] a very original technique to select an electron produced together with very many hadrons. The rejection power for an electron-positron pair could be as high as 10^{-6} against a background of π-pairs. For other hadrons the rejection was even more powerful. However, hadronic interactions were (and are) π-dominated, thus the new technique opened a new field of investigation: the possibility of observing (e^+e^-) pairs produced in hadronic interactions. This technique is now called «preshower» and is utilized whenever electrons need to be selected in a large hadronic background. For many years Zichichi was engaged at CERN in the study of lepton pair production, (e^+e^-), $(\mu^+\mu^-)$, in hadronic interactions. He studied the $(\bar{p}p)$ annihilation into lepton pairs [14], thus establishing the existence of a large timelike electromagnetic form factor for the nucleon.

He also studied (using the preshower method and another original device, the neutron missing-mass spectrometer), the (e^+e^-) decays of the heavy vector mesons and in particular the $(\omega-\phi)$ mixing [15]. When he came to Rome in order to discuss with me his project [1], he described the series of experiments performed at CERN during many years in lepton pair detection, in order to convince me that the technology he was proposing for the experimental set-up to be implemented in Frascati to search for a new heavy lepton had been well tested at CERN in several experiments.

I had also followed his work on the study of the $(e\mu)$ equivalence via the high-precision series of experiments on the anomalous magnetic moment of the muon. In fact, he was the inventor of the special technique to produce high-precision magnetic fields of polynomial shape. These special magnetic fields were needed in order, first to capture, then store and finally eject the muon from the magnet, so as to measure the spin rotation during its long storage in the magnet. The first paper of this series of high-precision work on the muon was published in 1960 [16]. The last paper was published in 1965 [17]. These five years of work show how much A. Zichichi was engaged in understanding the nature of the known leptons.

5 — Experimental work conducted by M.L. Perl during the years 1962-1973.

The research activity of the man who first observed acoplanar $(e\mu)$ pairs [5] is very different from that of Zichichi. Perl in the year 1962 published an interesting result on the shrinking of the Chew-Frautschi-Regge trajectories [18]. Contrary to what had been measured in the case of (pp) scattering, Perl and collaborators found that there was no shrinking of trajectories in (πp) scattering. Another interesting result [19] obtained by Perl was published in 1970 and refers to the measurement of neutron-proton elastic scattering. A further nice result published by Perl in 1973 [20] refers to the ratio of longitudinal-to-transverse polarization in ρ^0 electroproduction.

So there is no question that the interests of M.L. Perl were in hadron physics. But it is probably illuminating to study the sequence of events which took place in the years 1968-1971.

6 — Publications by M.L. Perl and the SLAC people during the years 1968-1971.

In 1968, M.L. Perl and collaborators published a paper [21] on *Search for New Particles Produced by High Energy Photons*. In this paper there is no mention of a Heavy Lepton with its own neutrino. This proves that no one at SLAC was thinking about this problem.

The first time that M.L. Perl discusses the problem of a heavy lepton of the type proposed and searched for in Frascati by A. Zichichi is in July 1971 [22]. The title of the paper is: *How does the muon differ from the electron?* and is published in Physics Today. Let me reproduce what M.L. Perl writes on page 35 of this paper: «*Fortunately these problems can be overcome in the newly developed electron-positron colliding-beam accelerators where leptons can be copiously produced through the process*[5] $e^+e^- \to \mu'^+\mu'^-$. *Within five years, through this process, we should know if the electron-muon family has additional members with masses in the GeV range.*» The reference (5) quoted by M.L. Perl is the first report published by A. Zichichi in 1970 [2] on the search for a heavy lepton with its own neutrino via the $(e\mu)$ acoplanar pairs in ADONE. A few months later, Y.S. Tsai published a paper [23] on *Decay correlations of Heavy Leptons in* $e^+e^- \to \ell^+\ell^-$. This is the first paper by SLAC people where a heavy lepton carrying its own leptonic number is discussed and studied.

In this paper [23], Tsai writes: «*Searches for these leptons have been attempted in the past*[1, 2]». Note that the reference (1) is the one we have quoted above [21] by A. Barna, ..., M.L. Perl et al. It is interesting to remark that Tsai adds (in the Reference part of the paper) following ref. (1), the following statement: «*Earlier attempts to search for heavy leptons can be traced from this paper*». This would mean that heavy leptons with their own neutrino were also included in ref. (1) quoted by Tsai, i.e. our ref. [21] quoted above. But of course there was no mention of such heavy leptons in this paper [21]. Reference (2) of Tsai's paper is the first published Frascati result [2] on the search for a heavy lepton with its own leptonic number via the acoplanar $(e\mu)$ method. In this Tsai paper [23], following ref. (2) there is also a reference to A.K. Mann (Lettere al Nuovo Cimento, 1, 486 (1971)), whose content is a wrong remark on the Frascati results [2]. The answer to this remark is in the paper containing the final Frascati result [3]. Obviously Tsai was carefully following the literature on the Frascati results concerning the Heavy Lepton.

The conclusion of this section is clear. M.L. Perl [22] read the Frascati paper [2] and realized its fundamental importance. Then he discussed with SLAC people and Tsai published a series of calculations [23] quoting an earlier SLAC paper [21] implying that this paper was also the original source of the new idea on the Sequential Heavy Lepton and on the method to search for it. This appears to me an attempt to lead one to believe that everything was done at SLAC. In this light it can be understood why in the first M.L. Perl publication [5] on the observation of acoplanar $(e\mu)$ pairs, the A. Zichichi work [2, 3] is never quoted in the references. Again, in the last review paper by M.L. Perl on *The discovery of the Tau Lepton* [24], one is given the impression that everything was done at SLAC. For example, although he now quotes the Frascati results, he does not mention the 1967 INFN proposal [1], nor the invention of the $(e\mu)$ method [1].

7 — Conclusions.

— As this historical review substantiates, the following is to Zichichi's credit:

- The idea for a new sequential Heavy Lepton.
- The invention of a new method to search for Heavy Leptons in $(e^+ e^-)$ annihilation via acoplanar $(e^{\pm}\mu^{\mp})$ pairs.
- The implementation of the new $(e^{\pm}\mu^{\mp})$ method.
- The proof that the method was working as expected in a real experiment.

— It is obvious that Perl picked up at once this idea, and took advantage of the existing MARK-II detector to discover the Heavy Lepton using the $(e^{\pm}\mu^{\mp})$ method invented by Zichichi.

On several occasions the original, fundamental contributions by Zichichi — who started this field — have been publicly acknowledged, first of all — even if not with consistency — by Perl himself, and by many others.

The discovery of the Heavy Lepton was made nearly two decades ago. During this time, in many scientific conferences, the authorship of the first idea, of the original $(e^{\pm}\mu^{\mp})$ method and of its implementation, has been attributed to A. Zichichi. I know of no other claims, in public or in private, to Zichichi's original idea (sequential Heavy Lepton) and/or to his invention of the acoplanar $(e^{\pm}\mu^{\mp})$ method that he has been able to implement and prove to work.

References.

[1]   *A proposal to search for Leptonic Quarks and Heavy Leptons produced at ADONE*
       M. Bernardini, D. Bollini, E. Fiorentino, F. Mainardi, T. Massam, L. Monari,
       F. Palmonari and A. Zichichi.
       INFN/AE-67/3, 20 March 1967.

[2]   *Limits on the electromagnetic production of Heavy Leptons*
       V. Alles-Borelli M. Bernardini, D. Bollini, P.L. Brunini, T. Massam, L. Monari,
       F. Palmonari and A. Zichichi.
       Lettere al Nuovo Cimento, 4, 1156 (1970).

[3]   *Limits on the mass of Heavy Leptons*
        M. Bernardini, D. Bollini, P.L. Brunini, E. Fiorentino, T. Massam, L. Monari,
        F. Palmonari, F. Rimondi and A. Zichichi.
        Nuovo Cimento, 17A, 383 (1973).

[4]   *Why (e⁺e⁻) Physics is fascinating*
        A. Zichichi.
        La Rivista del Nuovo Cimento, 4, 498 (1974), based on the contribution to many
        conferences, in particular the EPS Conference of Wiesbaden, 3-6 October 1972.

[5]   *Evidence for anomalous lepton production in e⁺-e⁻ annihilation*
        M.L. Perl et al.
        Physical Review Letters, 35, 1489 (1975).   In this paper, none of the above
        papers [1], [2], [3], [4] is quoted.

[6a]   *Validity of the leptonic selection rules for the ($\mu e \gamma$) vertex at high four-momentum
        transfers*
        V. Alles-Borelli, M. Bernardini, D. Bollini, P.L. Brunini, T. Massam, L. Monari,
        F. Palmonari and A. Zichichi.
        Lettere al Nuovo Cimento, 4, 1151 (1970) and

[6b]   *Study of charged final states produced in e⁺e⁻ interactions*
        V. Alles-Borelli, M. Bernardini, D. Bollini, P.L. Brunini, E. Fiorentino, T. Massam,
        L. Monari, F. Palmonari and A. Zichichi.
        Proceedings of the VIII Course of the «Ettore Majorana» International School of
        Subnuclear Physics, Erice, Italy, 1970: *«Elementary Processes at High Energy»*
        (Academic Press Inc., New York-London, 1971), p. 802.

[7]   *Experimental proof of the inadequacy of the peaking approximation in radiative
        corrections*
        V. Alles-Borelli, M. Bernardini, D. Bollini, P.L. Brunini, E. Fiorentino, T. Massam,
        L. Monari, F. Palmonari and A. Zichichi.
        Physics Letters, 36B, 149 (1971).

[8]   *Experimental check of crossing symmetry in the electromagnetic interaction of leptons*
        V. Alles-Borelli, M. Bernardini, D. Bollini, P.L. Brunini, E. Fiorentino, T. Massam,
        L. Monari, F. Palmonari and A. Zichichi.
        Lettere al Nuovo Cimento, 2, 376 (1971).

[9] *A check of quantum electrodynamics and of electron-muon equivalence*
V. Alles-Borelli, M. Bernardini, D. Bollini, P.L. Brunini, E. Fiorentino, T. Massam,
L. Monari, F. Palmonari and A. Zichichi.
Nuovo Cimento, 7A, 330 (1972).

[10] *Direct check of QED in e^+e^- interactions at high q^2 values*
V. Alles-Borelli, M. Bernardini, D. Bollini, P.L. Brunini, E. Fiorentino, T. Massam,
L. Monari, F. Palmonari and A. Zichichi.
Nuovo Cimento, 7A, 345 (1972).

[11] *An experimental study of acoplanar $(\mu^+\mu^-)$ pairs produced in (e^+e^-) annihilation*
D. Bollini, P. Giusti, T. Massam, L. Monari, F. Palmonari, G. Valenti and A. Zichichi.
Lettere al Nuovo Cimento, 13, 380 (1975).

[12] *Measurements of σ $(e^+e^- \rightarrow \mu^\pm\mu^\mp)$ in the energy range 1.2-3.0 GeV*
V. Alles-Borelli, M. Bernardini, D. Bollini, P. Giusti, T. Massam, L. Monari,
F. Palmonari, G. Valenti and A. Zichichi.
Physics Letters, 59B, 201 (1975).

[13] *A telescope to identify electrons in the presence of pion background*
T. Massam, Th. Muller and A. Zichichi.
CERN 63-25, 27 June 1963; Nuovo Cimento, 39, 464 (1965).

[14] *Discovery of the «Time-like structure of the proton» via the study of the leptonic annihilation modes of the proton-antiproton system*
M. Conversi, T. Massam, Th. Muller and A. Zichichi.
Physics Letters, 5, 195 (1963); Nuovo Cimento, 40, 690 (1965).

[15] *Observation of the rare decay modes $\omega \rightarrow e^+e^-$, $\phi \rightarrow e^+e^-$, and a direct determination of the $(\omega-\phi)$ mixing angle*
D. Bollini, A. Buhler-Broglin, P. Dalpiaz, T. Massam, F. Navach, F.L. Navarria,
M.A. Schneegans and A. Zichichi.
Nuovo Cimento, 56A, 1173 (1968); 57, 404 (1968).
See also the Review Paper:
The Basic SU$_3$ Mixing: $\omega_8 \leftrightarrow \omega_1$
A. Zichichi,
in *Evolution of Particle Physics* (Academic Press Inc., New York, N.Y., 1970),
p. 299.

[16] *A method for trapping muons in magnetic field*
G. Charpak, L.M. Lederman, J.C. Sens and A. Zichichi.
Nuovo Cimento, 17, 288 (1960).

[17]   *The anomalous magnetic moment of the muon*
G. Charpak, F.J. Farley, R.L. Garwin, Th. Muller, J.C. Sens and A. Zichichi.
Nuovo Cimento, 37, 1241 (1965).

[18]   *Pion-proton elastic diffraction scattering at 3, 4 and 5 GeV/c*
C.C. Ting, L.W. Jones and M.L. Perl.
Physical Review Letters, 9, 468 (1962).

[19]   *Neutron-proton elastic scattering from 8 to 30 GeV/c*
B.G. Gibbard, L.W. Jones, M.J. Longo, J. O'Fallon, J. Cox, M.L. Perl, W.T. Toner
and M.N. Kreisler.
Physical Review Letters, 24, 22 (1970).

[20]   *Electroproduction of ρ and φ mesons*
J.T. Dakin G.J. Feldman, W.L. Lakin, F. Martin, M.L. Perl, E.W. Petraske and
W.T. Toner.
Physical Review Letters, 30, 142 (1973).

[21]   *Search for New Particles Produced by High Energy Photons*
A. Barna, J. Cox, F. Martin, M.L. Perl, T.H. Tan, W.T. Toner, T.F. Zipf and
E.H. Bellamy.
Physical Review, 173, 1391 (1968). In this paper, the idea of a Heavy Lepton with its
own neutrino is missing.

[22]   *How does the muon differ from the electron?*
M.L. Perl.
Physics Today, July 1971. Here, the source of the new idea for a Heavy Lepton
carrying its own leptonic number is correctly quoted, i.e. Zichichi's work at Frascati
[2]. For unknown reasons, Perl calls this Heavy Lepton μ'.

[23]   *Decay correlations of Heavy Leptons in $e^+e^- \rightarrow \ell^+\ell^-$*
Y.S. Tsai.
Physical Review, 4D, 2821 (1971). In this paper, it is said that the idea for new Heavy
Leptons can be traced from ref. [21] above, as well as from ref. [2] above. This is
clearly incorrect because ref. [21] does not contain anything which concerns a Heavy
Lepton with its own neutrino.

[24]   *The Discovery of the Tau Lepton*
M.L. Perl.
SLAC-Pub-5937, September 1992 (T/E). Here, the Frascati results are quoted, but no
mention of ref. [1].

Marcello Conversi

CONCERNING THE DISCOVERY
OF THE HEAVY LEPTON

From

La Luce Pesante - Poligrafici Editoriale - Bologna, 1984 - p. 202

POLIGRAFICI EDITORIALE
BOLOGNA
1984

CONCERNING THE DISCOVERY OF THE HEAVY LEPTON

Marcello Conversi

« ... Concerning the discovery of the heavy lepton τ, the ADONE beam energy should have been 20% higher: in this case the τ would have been discovered at Frascati with the same method used later on at Stanford and first suggested, then tested at ADONE by Antonino Zichichi. ... »

André Petermann

THE ROOTS OF THE THIRD FAMILY

CNRS - Luminy - Marseille, France

THE ROOTS OF THE THIRD FAMILY

André Petermann

CNRS - Luminy - Marseille, France

ABSTRACT

The origin of the Third Family of leptons and quarks has its roots in the third lepton, with its own lepton conservation number. To open this new field, many years of work were needed with inventions and new technological developments. The present paper starts with my recollection of the events, in connection with the origin of the Third Family, which took place in CERN starting in the late 50's. It is followed by a detailed analysis of the reports presented by A. Zichichi and M. Perl at the International Conference on *The History of Original Ideas and Basic Discoveries in Particle Physics* and of the references quoted in their reports.

THE ROOTS OF THE THIRD FAMILY

André Petermann

1 — My recollection.

During the late fifties I did, for the first time, the theoretical computations of the anomalous magnetic moments of both the electron and the muon, and hence of the difference between them [1]. An experiment was being planned at CERN with the first proton accelerator, the SC. There were discussions about two possible options concerning the set-up: the "screw magnet" and the "flat magnet". The difficulty of realising a high-precision magnetic field using a flat magnet was overcome by A. Zichichi. He invented a very simple method to build high-precision fields of polynomial order (needed for injection, transition, storage and ejection of a muon beam) and this was the decisive step towards the high-precision measurement of the muon (g–2). Nino's interest in my theoretical calculations was for me a nice experience since physicists deeply involved in solving problems of technical nature would hardly care about the theoretical difficulties to be faced when performing detailed calculations. This is how we became friends. Everybody was hoping to get, as experimental value for the muon (g–2), a result in contrast with the theoretical predictions. This was not the case and the agreement between theory and experiment was established at the half of a percent level. This important experimental result, while confirming QED, did pose a new problem: that of electron-muon identity. Nino's approach to this now well-established identity was to search for heavier leptons carrying their leptonic number and being coupled to their own neutrinos. As Weisskopf recalls [2], this proposal by Nino was made before the discovery by Lederman, Schwartz and Steinberger of the two neutrinos $\nu_\mu \neq \nu_e$. The remarkable fact is that Nino did not limit himself to make a suggestion and forget it. He went on discussing with me and his other close friends his ideas on how to search for such a new lepton, emphasizing the technology that he planned to develop which would make the detection of acoplanar $e\mu$ pairs easy. For a new heavy lepton with its leptonic quantum number and with its own neutrino, the best production process was via a time-like photon. Nino decided to take the search for such a heavy lepton as a topic to be seriously investigated. For this search to become effective, an intense source of time-like photons was necessary. Hofstadter had established the existence of a space-like electromagnetic structure in the proton. What about the time-like region? Here again new interesting discussions started at CERN. Theoretically nothing could be predicted and the hypothesis of a point-like structure in the time-like region could not be excluded. If this had been the case, a high-intensity beam of antiprotons would have solved the problem of finding an intense source of time-like photons. But there was no such beam available at CERN or in other labs. Those were the times of strong interactions and bubble chamber physics. Zichichi succeeded in convincing Weisskopf (CERN DG at that time) to built the first high-intensity, partially separated beam of antiprotons. Unfortunately the time-like structure of the proton was far from being point-like and therefore the $\bar{p}p$ annihilation could not be the right production process for heavy lepton pairs.

It took many years to reach this conclusion and Nino got the green light by Weisskopf to use a new source of time-like photons for the production of heavy lepton (HL) pairs: the new e^+e^- collider being planned at Frascati. The fact that e^+e^- annihilation was going to be an intense source of (time-like) photons was on safe theoretical grounds and therefore, if HL had a point-like structure such as the known leptons, the production cross-section could be easily computed, even if a series of experimental checks were needed to establish the validity of QED in this range of q^2–values, never investigated before.

Once he had completed his series of searches at Frascati, Nino insisted in his engagement towards the search for a heavy lepton at higher energy. I remember his firm position that the Frascati energy level had to be increased as much as possible. In brief, if it were not for Nino's engagement, the search for a heavy lepton having as signature acoplanar $e\mu$ pairs would never have been started either at CERN or at Frascati.

When he presented the proposal (in the CERN Auditorium) to get support for his PAPLEP project (Proton AntiProton annihilation into LEpton Pairs) many objections were raised on the possibility of selecting electrons and muons with good efficiency and high rejection power against all sorts of background. The results obtained at CERN and Frascati showed that his acoplanar $e\mu$ method was working as predicted. It is exactly this method that led to the discovery of Zichichi's predicted heavy lepton, HL, now called τ lepton. My interest in this field is an old one and I was very sorry not to be able to attend the International Conference on "*The History of Original Ideas and Basic Discoveries in Particle Physics*".

In what follows I present a detailed analysis of the two reports presented at that Conference by A. Zichichi [3] and by M. Perl [4], respectively, and on the most relevant papers quoted by each author in order to back his own report.

2 — Analysis of Zichichi's Report.

The papers selected for the analysis of A. Zichichi's report [3], refer to the following topics:

i)   the invention of the "pre-shower" technique (1963) [5];

ii) the first large solid angle set-up for the detection of e^+e^- [6], $\mu^+\mu^-$ [7], $e^{\pm}\mu^{\mp}$ [8] pairs in hadronic interactions (1964);

iii) the INFN proposal to study via the acoplanar $e\mu$ method the production of a heavy lepton (HL) — now called "sequential" heavy lepton — carrying its own leptonic number and being coupled to its own neutrino (1967) [9]: in this proposal, for the first time in the literature, the "*most favourable production mechanisms and decay channels*" foreseen to search for a new sequential HL are rightly predicted;

iv) the validity of the $e \neq \mu$ leptonic selection rule at high q^2–values (1970) [10];

v) the first limit on the HL mass (1970) [11];

vi) the proof for the existence and the first measurement of acoplanar radiative effects (1971) [12];

vii) the final results on the HL mass limit (1973) [13].

From the analysis of these papers a clear sequence of events emerges. A sequence which is corroborated by the volume "*Lepton Physics at CERN and Frascati*", edited by N. Cabibbo [14]. Here is the result of our analysis.

In the late fifties thinking about leptons heavier than the heaviest lepton known at that time started at CERN [14] with A. Zichichi. His attempts developed towards the search for a one-GeV heavy lepton with its own leptonic number, thus being pair-produced and giving as signature an acoplanar $e^{\pm}\mu^{\mp}$ pair. The source of this new lepton was thought to be a time-like photon produced in $\bar{p}p$ hadronic interactions. The technology needed was devised at CERN [5], where the π/e "early-shower development" (now called "pre-shower") method was implemented in the first search for time-like photons in 1963 [6]. A large solid angle set-up, able to simultaneously detect $e^{\pm}e^{\mp}$, $\mu^{\pm}\mu^{\mp}$ and $e^{\pm}\mu^{\mp}$ pairs, was installed at the CERN PS and produced a series of experimental results: in 1964, on the time-like electromagnetic structure of the proton using the $\mu^{\pm}\mu^{\mp}$ channel [7]; in 1965, using both the $e^{\pm}e^{\mp}$ and $\mu^{\pm}\mu^{\mp}$ channels [8]; and later, on other physics problems connected with the production of time-like photons in hadronic interactions. Unfortunately hadronic interactions were not intense sources of time-like photons. As A. Zichichi says in his report [3] concerning the heavy lepton search, the conclusion was twofold: i) the $e\mu$–pair selection was indeed a very efficient tool to get rid of the overwhelming background; ii) the primary production process for the time-like photons could not be $\bar{p}p$.

The π/e and π/μ identification technique, together with the idea to search for a heavy lepton carrying its own leptonic number and being pair-produced by a powerful source of time-like photons, the ADONE $e^{+}e^{-}$ collider, was brought by A. Zichichi from CERN to Frascati [9] where, in 1970, the first limit on the heavy lepton mass was obtained [11]. The 1970 publication by the Bologna-CERN-Frascati group stimulated a lot of interest in the scientific community. In fact, contrary to most people's wisdom, the search for a sequential heavy lepton turned out to be indeed possible and the acoplanar $e^{\pm}\mu^{\mp}$ method the best and cleanest way to look for it.

3 — Analysis of Perl's Report.

Let us now turn to the analysis of M. Perl's report [4] and of the most relevant papers quoted therein.

The publications quoted by M. Perl can be classified as follows.

— Those published before 1970 (i.e. before the first Frascati result by A. Zichichi et al.), with two papers published before 1967 [9], the date of the original Frascati proposal and a third one published after 1967. The first two are: i) the 1961 paper by Cabibbo and Gatto [15], defined "*seminal*" by M. Perl, despite the fact that this paper has nothing which refers to a heavy lepton of sequential nature; ii) the 1964 paper by E.M. Lipmanov [16]: here the heavy lepton is μ'^+ with the same lepton number as the negative muon; the third one is the 1969 paper by Rothe and Wolsky [17] which, if right, would have not allowed[*] Perl to discover any "*anomalous lepton production in e^+e^- annihilation*" [18]. In all the papers quoted by Perl and published before the 1970 Frascati result, either the idea of a new sequential heavy lepton, or the prediction of the correct experimental way to detect it, is missing, or both.

— Those published after 1970: in these papers the concept of a heavy lepton of sequential nature is indeed present, regarding both theoretical calculations (even if elementary [19]), and detection possibilities. But all these papers follow the first Frascati result of 1970 [11]. Therefore they cannot be considered as original contributions to the idea of a sequential heavy lepton, nor to the method (acoplanar $e\mu$ final states) and technologies needed to detect such a new type of lepton.

In our opinion, the most relevant papers quoted by M. Perl are the following:

i) *Search for new particles produced by high energy photons*, published by A. Barna et al. in 1968 [20];

ii) *Decay correlations of heavy leptons in $e^+e^- \rightarrow l^+l^-$*, published by Y.S. Tsai in 1971 [19];

iii) *Evidence for anomalous lepton production in e^+e^- annihilation*, published by M. Perl et al. in 1975 [18]: this is the paper by M. Perl on the discovery of acoplanar $e\mu$ events produced in e^+e^- annihilation.

iv) *How does the muon differ from the electron?*, published by M. Perl in Physics Today, in 1971 [21].

In this paper Perl wrote:

"*Are the muon and the electron part of a larger family of charged leptons?*" ...

"*Fortunately these problems can be overcome in the newly developed electron-*

(*) In fact Rothe's and Wolsky's work on a theoretically wrong model for weak interactions (SU(2)) had as conclusion the search for acoplanar pion pairs in the final state: the acoplanar $e\mu$ pairs were not mentioned.

positron colliding-beam accelerators where charged leptons can be copiously produced through the process[5]

$$e^+ + e^- \rightarrow \mu'^+ + \mu'^-$$

Within five years, through this process we shall know if the electron-muon family has additional members with masses in the several-GeV ranges."

Reference 5 (shown as superscript) in this paper refers to the 1970 result by A. Zichichi et al. [11].

From the study of Perl's report [4] and the references mentioned above, the following points emerge.

1) The 1975 paper where the acoplanar $e^{\pm}\mu^{\mp}$ pairs were first observed by M.L. Perl et al. [18] (*Evidence for anomalous lepton production in e^+e^- annihilation*) does not contain any reference to the Frascati work by A. Zichichi and collaborators: neither to the 1967 INFN proposal [9]; nor to the first (1970) and final (1973) mass limits on a sequential heavy lepton searched for by means of acoplanar $e^{\pm}\mu^{\mp}$ pairs [11, 13]; nor to the first detection of acoplanar radiative effects [12] and the sequence of QED checks [14]; nor to the work on the validity of the $e \neq \mu$ leptonic selection rule at high q^2–values [10]. All these results were published by the Bologna-CERN-Frascati group before the discovery of acoplanar $e\mu$ pairs at SLAC.

2) The (July 1971) Physics Today paper by M. Perl [21] quotes as Ref. 5 the 1970 first limit on the heavy lepton mass obtained at ADONE by Zichichi's group [11]. This paper [11] was therefore known to M. Perl when, in 1975, he published the first evidence for acoplanar $e^{\pm}\mu^{\mp}$ pairs produced in e^+e^- annihilation at SPEAR [18]. Actually, going through the various papers by M. Perl (et al.) quoted by himself in his report [4], one cannot but observe a gap concerning the years 1975-1976, where the results by A. Zichichi et al. [11, 13] cease to be duly referenced.

3) The 1971 paper by Y.S. Tsai [19] (*Decay correlations of heavy leptons in $e^+e^- \rightarrow l^+l^-$*), quoted by M. Perl in his report at the Conference [4], contains a statement: "*Searches for these leptons have been attempted in the past*" (quoting Ref. 1) and in the reference section of the same paper Tsai adds, concerning Ref. 1: "*Earlier attempts to search for heavy leptons can be traced from this paper*". The paper quoted by Tsai as Ref. 1 is the already mentioned 1968 work by Perl et al. [20] (A. Barna et al., *Search for new particles produced by high energy photons*). But this paper does not have anything which refers, either directly or even indirectly, to a heavy lepton carrying its own leptonic number and being coupled to its own neutrino.

4) M. Perl says in his report [4] that since the early 1960's he was thinking about sequential heavy leptons and $e\mu$ signals. If M. Perl had been thinking for so many years on how to detect such a new kind of heavy lepton, why were the detection systems of MARK I so "*crude*" and "*just well enough*" for electrons and muons?

5) M. Perl says in his report [4]: "*It was this 1964 proposal and the 1961 seminal paper of Cabibbo and Gatto ... which focussed my thinking on new charged lepton searches using an e^+e^- collider*". As already pointed out, the Cabibbo and Gatto paper [15] has nothing on sequential heavy leptons. The 1964 SLAC proposal is unpublished and we could not find it in any library. Hence we could not read it and we leave to the reader the task to discover if the heavy lepton and the acoplanar $e\mu$ method are discussed therein. In any case the SLAC proposal was not even mentioned in the 1971 paper by Perl [21] published in Physics Today and quoted in point 2) above. It is also peculiar that the 1961 "*seminal*" paper by Cabibbo and Gatto was not mentioned in this Physics Today paper [21], nor in any other publication by Perl among all those he quotes in his report [4].

4 — Conclusions.

The analysis of the two reports at the Conference [3, 4] and of the papers [5 - 26] allow us to reach the following conclusions: Zichichi's experimental team first had the idea of the possible existence of a heavy lepton, $HL^{\pm}$, with its own neutrino, i.e. carrying its own leptonic quantum number and invented the acoplanar $e\mu$ method to search for such a lepton. Furthermore they built the set-up able to experimentally implement this $e\mu$ method and established that the best process to search for $HL^{\pm}$–pair production was e^+e^- annihilation. Finally, they proved, by performing the first dedicated experiment in e^+e^- annihilation, that if $HL^{\pm}$–pair production had taken place, the background would have been almost nulled and the acoplanar $e\mu$ signal clearly observed.

These points are the compelling way to the Third Family of leptons and quarks.

The discovery by M. Perl of the $e\mu$ signal, proposed and searched for by A. Zichichi over more than a decade, appears as a logical consequence of executive nature. The main original ideas (such as that of a new sequential heavy lepton and of the acoplanar $e\mu$ signal to search for it in e^+e^-), and the adequate experimental set-up had been put forward years before Perl's effective discoveries, as established by the published records, and in accordance with my personal recollection.

References.

[1]   *Magnetic Moment of the μ Meson*
        A. Petermann.
        Physical Review, 105, 1931 (1957).

        Fourth order magnetic moment of the electron
        A. Petermann.
        Helvetica Physica Acta, 30, 407 (1957).

[2]   *The heartbeat of the proton*
        V.F. Weisskopf.
        in *"Lepton Physics at CERN and Frascati"*, N. Cabibbo Ed., 20th Century Physics
        Series, Vol. 8, World Scientific, 45 (1994).

[3]   *Foundations of sequential Heavy Lepton searches*
        A. Zichichi.
        Presented at the International Conference on "The History of Original Ideas and Basic
        Discoveries in Particles Physics", 1994, H.B. Newman and T. Ypsilantis Eds., Plenum
        Press (1996), Vol. 352, 227.

[4]   *The discovery of the tau lepton. Part 1: The early history through 1975*
        M. Perl.
        Presented at the International Conference on "The History of Original Ideas and Basic
        Discoveries in Particles Physics", 1994, H.B. Newman and T. Ypsilantis Eds., Plenum
        Press (1996), Vol. 352, 277.

[5]   *A telescope to identify electrons in the presence of pion background*
        T. Massam, Th. Muller and A. Zichichi.
        CERN 63-25, 27 June 1963.

[6]   *Search for the timelike structure of the proton*
        M. Conversi, T. Massam, Th. Muller and A. Zichichi.
        Physics Letters, 5, 195 (1963).

[7]   *Proton antiproton annihilation into muon pair*
        M. Conversi, T. Massam, Th. Muller, M. Schneegans and A. Zichichi.
        Proceedings of the International Conference on "High-Energy Physics", Dubna, USSR,
        5-15 August 1964 (Atomizdat, Moscow, 1966), Vol. I, 857.

[8]   *The leptonic annihilation modes of the proton-antiproton system at 6.8 (GeV/c)² timelike*
        four-momentum transfer
        M. Conversi, T. Massam, Th. Muller and A. Zichichi.
        Nuovo Cimento, 40, 690 (1965).

[9]   *A proposal to search for leptonic quarks and Heavy Leptons produced by ADONE*
      M. Bernardini, D. Bollini, E. Fiorentino, F. Mainardi, T. Massam, L. Monari,
      F. Palmonari and A. Zichichi.
      INFN/AE-67/3, 20 March 1967.

[10] *Validity of the leptonic selection rules for the (μeγ) vertex at high four-momentum*
      transfers
      V. Alles-Borelli, M. Bernardini, D. Bollini, P.L. Brunini, T. Massam, L. Monari,
      F. Palmonari and A. Zichichi.
      Lettere al Nuovo Cimento, 4, 1151 (1970).

[11] *Limits on the electromagnetic production of Heavy Leptons*
      V. Alles-Borelli, M. Bernardini, D. Bollini, P.L. Brunini, T. Massam, L. Monari,
      F. Palmonari and A. Zichichi.
      Lettere al Nuovo Cimento, 4, 1156 (1970).

[12] *Experimental proof of the inadequacy of the peaking approximation in radiative*
      corrections
      V. Alles-Borelli, M. Bernardini, D. Bollini, P.L. Brunini, E. Fiorentino, T. Massam,
      L. Monari, F. Palmonari and A. Zichichi.
      Physics Letters, 36B, 149 (1971).

[13] *Limits on the mass of Heavy Leptons*
      P.L. Brunini, E. Fiorentino, T. Massam, L. Monari, F. Palmonari, F. Rimondi and
      A. Zichichi.
      Nuovo Cimento, 17A, 383 (1973).

[14] *Lepton physics at CERN and Frascati*
      N. Cabibbo Ed., 20th Century Physics Series, Vol. 8, World Scientific (1994).

[15] *Electron-positron colliding beam experiments*
      N. Cabibbo and R. Gatto.
      Physical Review, 124, 1577 (1961).

[16] *The Question of the Possible Existence of a Heavy Charged Lepton*
      E.M. Lipmanov.
      Soviet Physics JETP, 19, 1291 (1964).

[17] *Are there heavy leptons?*
      K.W. Rothe and A.M. Wolsky.
      Nuclear Physics, B10, 241 (1969).

[18] *Evidence for anomalous lepton production in e^+e^- annihilation*
      M. Perl et al.
      Physical Review Letters, 35, 1489 (1975).

[19] *Decay correlations of Heavy Leptons in $e^+e^- \to l^+l^-$*
      Y.S. Tsai.
      Physical Review, D4, 2821 (1971).

[20] *Search for new particles produced by high energy photons*
      A. Barna, J. Cox, F. Martin, M.L. Perl, T.H. Tan, W.T. Toner, T.F. Zipf and
      E.H. Bellamy.
      Physical Review, 173, 1391 (1968).

[21] *How does the muon differ from the electron?*
      M. Perl.
      Physics Today, July 1971, 34.

[22] *Foreword*
      N. Cabibbo.
      in "*Lepton Physics at CERN and Frascati*", N. Cabibbo Ed., 20th Century Physics
      Series, Vol. 8, World Scientific, xiii (1994).

[23] *Study of charged final states produced in e^+e^- interactions*
      V. Alles-Borelli, M. Bernardini, D. Bollini, P.L. Brunini, E. Fiorentino, T. Massam,
      L. Monari, F. Palmonari and A. Zichichi.
      Proceedings of the VIII Course of the "Ettore Majorana" International School of
      Subnuclear Physics, Erice, Italy, 1970: "Elementary Processes at High Energy"
      (Academic Press Inc., New York-London, 1971), 790.

[24] *Proof of hadron production in e^+e^- interactions*
      V. Alles-Borelli, M. Bernardini, D. Bollini, P.L. Brunini, F. Fiorentino, T. Massam,
      L. Monari, F. Palmonari, G. Valenti and A. Zichichi.
      Proceedings of the International Conference on "Meson Resonances and Related
      Electromagnetic Phenomena", Bologna, Italy, 14-16 April 1971 (Editrice Compositori,
      Bologna, 1972), 489.

[25] *A study of the hadronic angular distribution in (e^+e^-) processes from 1.2 to 3.0 GeV*
      M. Bernardini, D. Bollini, P.L. Brunini, F. Fiorentino, T. Massam, L. Monari,
      F. Palmonari, F. Rimondi and A. Zichichi.
      Nuovo Cimento, 26A, 163 (1975).

[26] *Acoplanar (e^+e^-) pairs and radiative corrections*
      M. Bernardini, D. Bollini, P.L. Brunini, E. Fiorentino, T. Massam, L. Monari,
      F. Palmonari, F. Rimondi and A. Zichichi.
      Physics Letters, 45B, 169 (1973).

Björn H. Wiik and Günter Wolf

THE SEARCH FOR HEAVY LEPTONS BY A. ZICHICHI AND HIS COLLABORATORS

Deutsches Elektronen Synchrotron DESY

THE SEARCH FOR HEAVY LEPTONS BY A. ZICHICHI
AND HIS COLLABORATORS

Björn H. Wiik and Günter Wolf
Deutsches Elektronen Synchrotron DESY

A. Zichichi and his group have set out to study $p\bar{p} \rightarrow ee$, $\mu\mu$ starting 1960. As V. Weisskopf recalls, the underlying aim of this research was to establish $p\bar{p}$ scattering as a source of massive photons from annihilation ($p\bar{p} \rightarrow \gamma^*$) so that $p\bar{p}$ scattering could be used for the search of a new heavy lepton.

In order to achieve its goals the group developed, during several years of research, the preshower technique for separating electrons from hadrons with high degree of certainty (see [1, 2, 3]) and to improve the identification of muons (see [4]). The methods developed proved to be very successful. Thanks to the new techniques the group, working at CERN, was able to put an upper limit on the transition rate for

$$p\bar{p} \rightarrow e^+e^-, \quad \mu^+\mu^-$$

compared to the total $p\bar{p}$ annihilation rate as low as $1.1 \cdot 10^{-8}$ (see [5, 6]). The measured large suppression of the leptonic modes indicated a strong formfactor of the nucleon and made it clear that $p\bar{p}$ was not an intense source of highly virtual timelike photons and therefore not a good place to search for heavy leptons.

The group demonstrated the power of their detector by the observation, at CERN, of the leptonic decay of the ϕ and ω mesons studying the reactions

$$\pi^- p \rightarrow n\phi, \qquad \phi \rightarrow e^+e^-$$

and

$$\pi^- p \rightarrow n\omega, \qquad \omega \rightarrow e^+e^-$$

(see [7, 8]). The experiment yielded the first measurements of the partial widths

$$\Gamma_{\phi \rightarrow e^+e^-} = (2.1 \pm 0.9)\,\text{keV}$$

and

$$\Gamma_{\omega \rightarrow e^+e^-} = (0.49 \pm 0.19)\,\text{keV}$$

which are in good agreement with the best values of today; viz. the 1996 PDG values [9] are $(1.37 \pm 0.05)\,\text{keV}$ and $(0.60 \pm 0.02)\,\text{keV}$, respectively. In the case of the ϕ, the experiment succeeded to isolate the $\phi \rightarrow e^+e^-$ decays from the hadronic processes which were one million times more copious.

In 1967 the group (now called the BCF group) submitted a proposal to the Frascati Laboratory entitled: A Proposal to Search for Leptonic Quarks and Heavy Leptons at ADONE (see [10]). The paper discussed, amongst other topics, a proposal to search for heavy leptons produced by e^+e^- annihilation. These hypothetical heavy leptons $HL^{\pm}$ were assumed to carry a new lepton quantum number. As a result, they would decay via

$$HL \rightarrow e\nu_L\overline{\nu_e}$$

and

$$HL \rightarrow \mu\nu_L\overline{\nu_\mu} \; .$$

The best way of searching for pair production of HL would be to look for events with nothing but acoplanar e and μ observed in the final state plus missing momentum. The proposed detector was based on the techniques for e / hadron and μ / hadron separation developed by the group in their previous experiments which had reached a π / e suppression of $5 \cdot 10^{-4}$, (see Figs. 1[a, b, c] [2, 3]) for momenta up to 2.5 GeV / c. The π / μ suppression was decreasing from $1.8 \cdot 10^{-2}$ at 1.25 GeV / c to $1.5 \cdot 10^{-3}$ at 2.5 GeV / c (see Fig. 2 and Table I [6]).

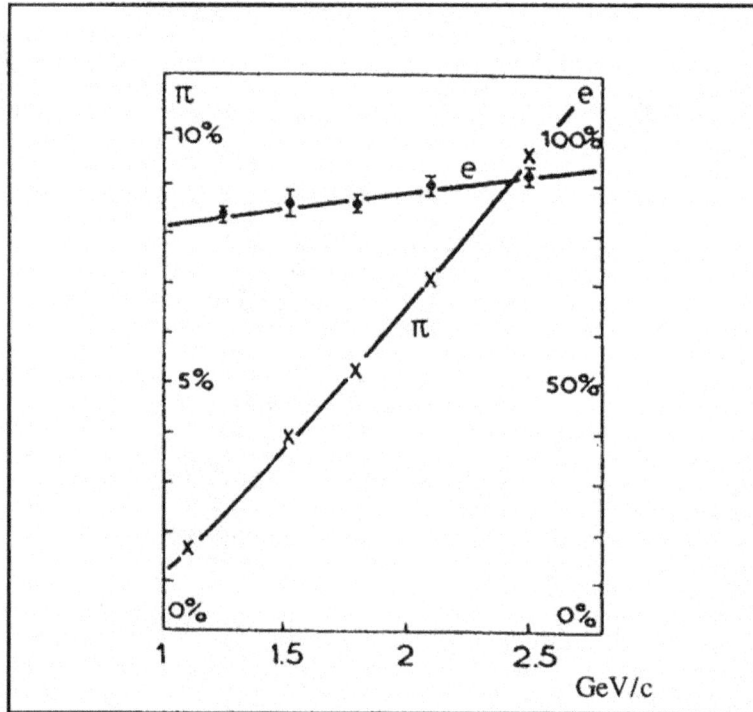

Fig. 1 a: Electron-pion separation in the momentum range: $(1 \div 2.5)$ GeV/c. Variation with energy of the electronic efficiency of pions and electrons.

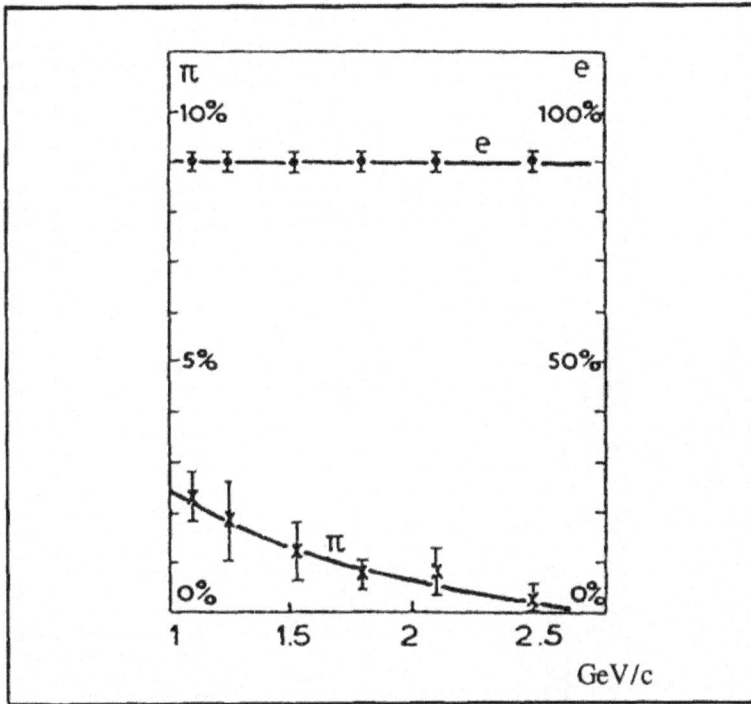

Fig. 1 b: Electron-pion separation in the momentum range: $(1 \div 2.5)$ GeV/c. Variation with energy of the fraction of electronically selected pion events which are also accepted as electrons in the spark-chamber analysis.

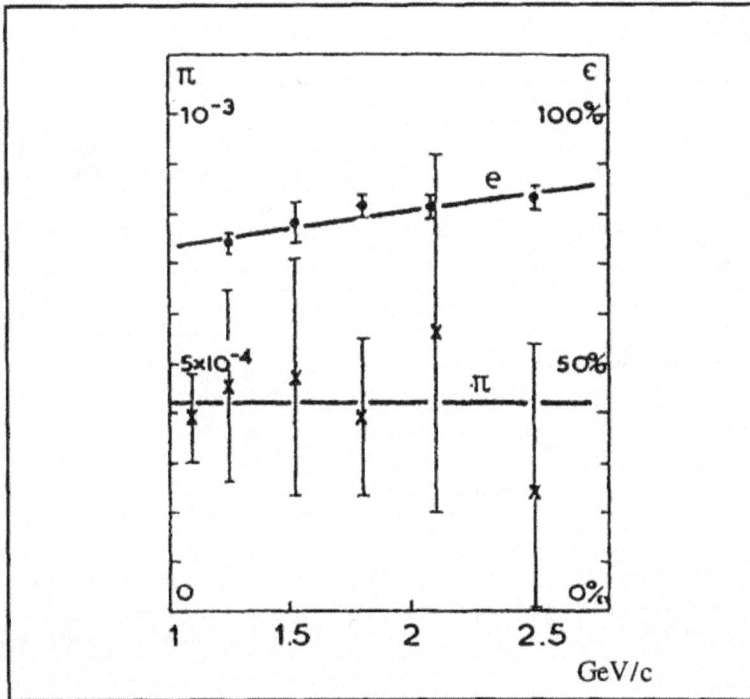

Fig. 1 c: Electron-pion separation in the momentum range: $(1 \div 2.5)$ GeV/c. Over-all efficiency for pions and electrons as a function of energy.

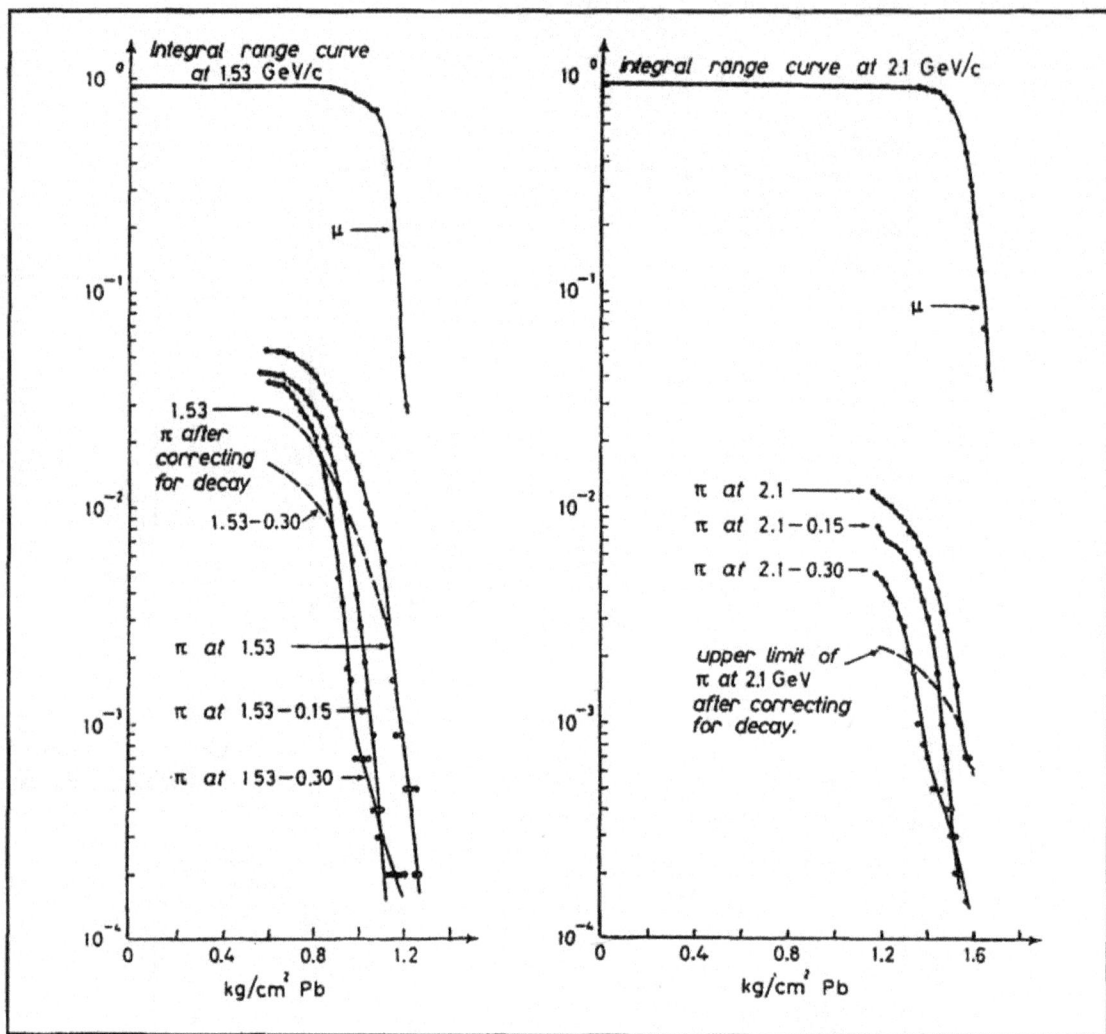

Fig. 2: Range distributions of muons and pions: (left) for muons of 1.53 GeV / c and pions of 1.53, 1.38 and 1.23 GeV / c; (right) for muons of 2.10 GeV / c and pions of 2.10, 1.95 and 1.80 GeV / c.

Energy in GeV	Penetrations
2.5	$1.5 \cdot 10^{-3}$
2.1	$1.0 \cdot 10^{-3}$
1.8	$6.0 \cdot 10^{-3}$
1.5	$1.5 \cdot 10^{-2}$
1.25	$1.8 \cdot 10^{-2}$

Table I: Typical values for the pion penetration which corresponded to 90% muon efficiency.

The proposal was approved and the detector constructed. From the beginning of collider operation the group concentrated on leptonic processes. The measured cross sections for collinear and noncollinear e^+e^- and $\mu^+\mu^-$ pair production showed agreement with the expectations from QED in the peaking approximation. The group observed for the first time acoplanar e^+e^- (see Fig. 3 [12]) and $\mu^+\mu^-$ events (see Fig. 4 [13]) and showed that they also agreed with QED provided one calculated the full radiative effects instead of using the peaking approximation (see [11, 12, 13]).

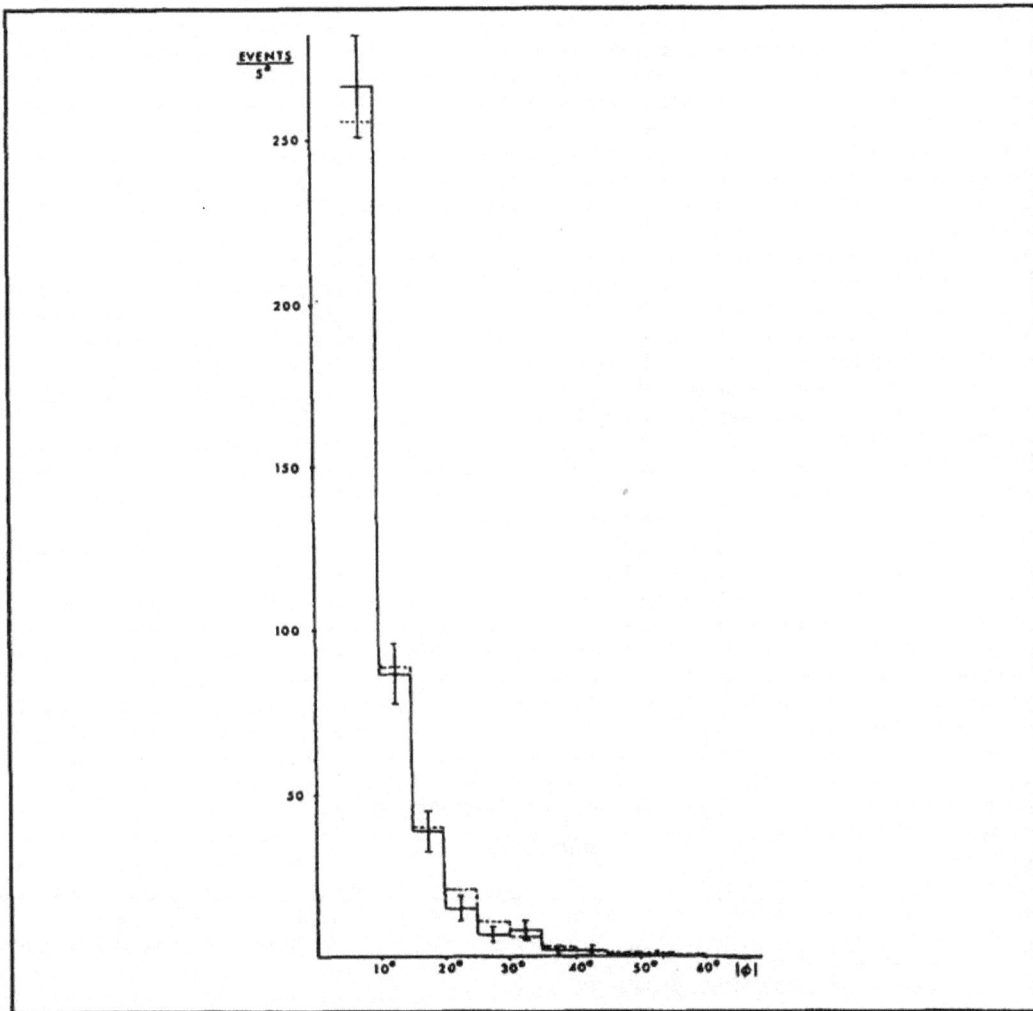

Fig. 3: Acoplanarity distribution for e^+e^- pairs. The full histogram with error bars shows the data; the dashed histogram shows the QED prediction.

The agreement of acoplanar e^+e^- and $\mu^+\mu^-$ events with QED was an important test in the search for heavy leptons where the BCF group was looking for acoplanar $e\mu$ events. The measurements gave also proof that the group understood quantitatively the performance of the detector, including its acceptance and its efficiency for identifying electrons and muons.

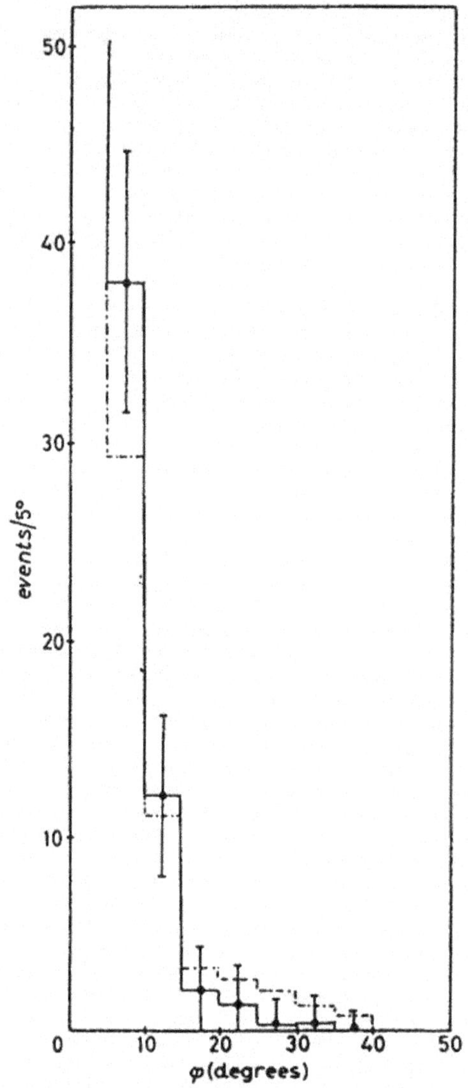

Fig. 4:   Acoplanarity distribution for $\mu^+\mu^-$ pairs. The full histogram with error bars shows the data; the dashed histogram shows the QED prediction.

The group then undertook a massive search for heavy leptons scanning the beam energy range from E = 0.6 to 1.5 GeV, 1.5 GeV being then maximum energy accessible at ADONE (see [14, 15]). With a total luminosity of about 450 nb^{-1} only two acoplanar $e\mu$ event candidates were found, which ruled out the existence of heavy leptons with a mass below 1 Gev. Again, the techniques used for identifying electrons and muons were the key to this result: only two background events were found in the presence of a total of about 19'000 hadronic events produced over the same running period. The small background demonstrated the feasibility of the $e\mu$ acoplanarity method for the search for heavy leptons and the strength of the detector with respect to e/π and μ/π separation.

As we know today, the τ has a mass of 1.777 GeV. The discovery of the τ by the BCF group would have required an upgrade of the beam energy at ADONE from 1.5 to 2.0 GeV and an increase of the solid angle covered by the BCF detector from $\Delta\Omega/4\pi \approx 0.16$ to about 0.5. The increase in solid angle by a factor of three would have increased the acceptance for $e\mu$ events from $\tau\bar{\tau}$ production by a factor of $3^3 = 9$. The envisaged methodology, namely the search for acoplanar $e\mu$ events would have been well suited and the e/π and μ/π separation capabilities of the detector would have allowed to discover the τ. Presumably, the BCF group would have continued to search for the τ in the same way as before, namely by increasing the beam energy of ADONE in steps of 0.1 GeV and collecting at each energy a substantial amount of luminosity. With about 400 nb^{-1} per energy setting no signal would have been observed at $E_{beam} \leq 1.7$ GeV while approximately 10, 19 and 20 signal events from $\tau\bar{\tau}$ production would have been found at $E_{beam} = 1.8$, 1.9 and 2.0 GeV, respectively.

References.

[1]   *A Telescope to Identify Electrons in the Presence of Pion Background*
       T. Massam, Th. Muller and A. Zichichi.
       CERN 63-25 (1963).

[2]   *A New Electron Detector with High Rejection Power against Pions*
       T. Massam, Th. Muller, M.A. Schneegans and A. Zichichi.
       Nuovo Cimento, 39, 464 (1965).

[3]   *A Large Electromagnetic Shower Detector with High Rejection Power against Pions*
       M. Basile, J. Berbiers, D. Bollini, A. Buhler-Broglin, P. Dalpiaz, F.L. Frabetti,
       T. Massam, F. Navach, F.L. Navarria, M.A. Schneegans and A. Zichichi.
       Nuclear Instruments and Methods, 101, 433 (1972).

[4]   *Range Measurements for Muons in the GeV Region*
       A. Buhler, T. Massam, Th. Muller and A. Zichichi.
       Nuovo Cimento, 35, 759 (1965).

[5]   *Search for the Time-Like Structure of the Proton*
       M. Conversi, T. Massam, Th. Muller and A. Zichichi.
       Physics Letters, 5, 195 (1963).

[6]   *The Leptonic Annihilation Modes of the Proton-Antiproton System at* $6.8 \ (GeV/c)^2$
      Timelike Four - Momentum Transfer
      M. Conversi, T. Massam, Th. Muller and A. Zichichi.
      Nuovo Cimento, 40, 690 (1965).

[7]   *Observation of the Rare Decay Mode of the* ϕ-*Meson:* $\phi \rightarrow e^+ e^-$
      D. Bollini, A. Buhler-Broglin, P. Dalpiaz, T. Massam, F. Navach, F.L. Navarria,
      M.A. Schneegans and A. Zichichi.
      Nuovo Cimento, 56A, 1173 (1968).

[8]   *The Decay Mode* $\omega \rightarrow e^+ e^-$ *and a Direct Determination of the* $\omega - \phi$ *Mixing Angle*
      D. Bollini, A. Buhler-Broglin, P. Dalpiaz, T. Massam, F. Navach, F.L. Navarria,
      M.A. Schneegans and A. Zichichi.
      Nuovo Cimento, 57A, 404 (1968).

[9]   *Review of Particle Physics*
      R. M. Barnett et al.
      Physical Review, D54, 1 (1996).

[10] *A Proposal to Search for Leptonic Quarks and Heavy Leptons Produced by ADONE*
      M. Bernardini, D. Bollini, E. Fiorentino, F. Mainardi, T. Massam, L. Monari,
      F. Palmonari and A. Zichichi.
      INFN/AE-67/3 (1967).

[11] *Experimental Proof of the Inadequacy of the Peaking Approximation in Radiative*
      Corrections
      V. Alles-Borelli, M. Bernardini, D. Bollini, P.L. Brunini, E. Fiorentino, T. Massam,
      L. Monari, F. Palmonari and A. Zichichi.
      Physics Letters, 36B, 149 (1971).

[12] *Acoplanar* $(e^+ e^-)$ *Pairs and Radiative Corrections*
      M. Bernardini, D. Bollini, P.L. Brunini, E. Fiorentino, T. Massam, L. Monari,
      F. Palmonari, F. Rimondi and A. Zichichi.
      Physics Letters, 45B, 169 (1973).

[13] *An Experimental Study of Acoplanar* $(\mu^{\pm} \mu^{\mp})$ *Pairs Produced in* $(e^+ e^-)$ *Annihilation*
      D. Bollini, P. Giusti, T. Massam, L. Monari, F. Palmonari, G. Valenti and
      A. Zichichi.
      Lettere al Nuovo Cimento, 13, 380 (1975).

[14] *Limits on the Electromagnetic Production of Heavy Leptons*
      V. Alles-Borelli, M. Bernardini, D. Bollini, P.L. Brunini, T. Massam, L. Monari,
      F. Palmonari and A. Zichichi.
      Lettere al Nuovo Cimento, 4, 1156 (1970).

[15] *Limits on the Mass of Heavy Leptons*
      M. Bernardini, D. Bollini, P.L. Brunini, E. Fiorentino, T. Massam, L. Monari,
      F. Palmonari, F. Rimondi and A. Zichichi.
      Nuovo Cimento, 17A, 383 (1973).

Appendices A and B

Antonino Zichichi

FOUNDATIONS OF
SEQUENTIAL HEAVY LEPTON SEARCHES

From
The History of Original Ideas and Basic Discoveries in Particle Physics
Edited by H.B. Newman and T. Ypsilantis - 1996 - pp. 227-273.

PLENUM PUBLISHING CORPORATION
NEW YORK
1996

FOUNDATIONS OF SEQUENTIAL HEAVY LEPTON SEARCHES

Antonino Zichichi

CERN, Geneva, Switzerland

ABSTRACT

The original contributions of the Bologna-CERN-Frascati (BCF) group to the discovery — via the acoplanar $(e^{\pm}\mu^{\mp})$ method — of the Heavy Lepton (HL, now called τ) are described. These contributions include: i) the idea of a new lepton in the GeV mass range, carrying its own leptonic number; ii) the search and the identification of the best production process: $e^+e^- \rightarrow HL^+HL^-$; iii) the invention of the acoplanar $(e^{\pm}\mu^{\mp})$ method with the associated technology and the proof that it works; iv) the implementation of the large solid-angle detector needed to establish the first upper limit on the HL mass at Frascati; v) the promotion of the HL searches at energies higher than ADONE. These original contributions started in 1960 at CERN where the search was based on the time-like photons produced in $(\bar{p}p)$ annihilation. This search led to the construction of the first large solid-angle detector able to simultaneously detect electrons and muons with high rejection power against pions, thus allowing, already in 1964, to establish the validity of the $(e^{\pm}\mu^{\mp})$ method as the best one to detect Heavy Lepton pair production. On the other hand the CERN experiment (PAPLEP) led to the conclusion that the best source of time-like photons had to be (e^+e^-) since the proton was found to have a strong time-like electromagnetic structure.

The search for a Heavy Lepton (HL), first at CERN (1960-1964) then at Frascati (1967-1973), using innovative technologies, represents more than ten years of work performed before the SLAC discovery in 1975. The report is a synthesis of the events, validated by the references to the original works and by their time sequence.

From this analysis it turns out that, prior to 1967 (the original Frascati proposal), there is not a single publication, technological invention or experimental work which could be linked to the existence of, and the search for, a Heavy Lepton (HL) carrying its own leptonic number, and coupled to its own neutrino: the symbol adopted was HL and corresponds to what is now known as the "sequential" heavy lepton τ. This term "sequential" was coined, not before 1967, but after the more than ten years of work (1960-1973) carried on at CERN and Frascati. The actual discovery of the third lepton probably would not have occurred the way it did if M. Perl had not read the first Frascati results, published in 1970, where the new ideas on HL and on its best signature, i.e. acoplanar $(e\mu)$ pairs, were described. The acoplanar $(e\mu)$ pairs from time-like photons were predicted to be the best signature for $HL^{\pm}$ pair production in (e^+e^-) annihilation, and this was indeed the meaning of the "anomalous lepton production" discovered by Perl at SLAC in 1975.

Presented at the International Conference on:
"The History of Original Ideas and Basic Discoveries in Particle Physics"

FOUNDATIONS OF SEQUENTIAL HEAVY LEPTON SEARCHES

Antonino Zichichi

Table of Contents

1 — Introduction.

The Heavy Lepton (HL) was discovered (and called τ) by Martin Perl et al. [1] in 1975, via the study of the production process:

$$e^+e^- \rightarrow HL^+ HL^- \tag{1}$$

and the decay channels:

$$HL^\pm \underbrace{\qquad}_{\begin{array}{l} \rightarrow e^\pm + \nu_e + \nu_{HL} \\ \rightarrow \mu^\pm + \nu_\mu + \nu_{HL} \end{array}} \tag{2}$$

using the method of detecting acoplanar $(e^\pm \mu^\mp)$ pairs. Reactions (1) and (2) are on page 7 of the INFN proposal, dated 1967, to search for Heavy Leptons with the ADONE (e^+e^-) collider at Frascati [2]. Figure 1.1 is the front page of the proposal and Figure 1.2 the page where the production reaction (1) and the decay channels (2) are given.

The acoplanar $(e^\pm \mu^\mp)$ method was the key point of the Bologna-CERN-Frascati (BCF) proposal to INFN, which is, in turn, based on the fundamental idea that there is a new leptonic number associated with the Heavy Lepton.

As we will see in this report, thinking on how to search for a Heavy Lepton carrying its own leptonic number started at CERN in the late fifties during the first high precision measurements of the muon (g–2) [3]. This thinking led in 1963 to the invention of the "early-shower development" method, now called "pre-shower" method [4-7], and to a series of experiments on lepton pair studies in hadronic interactions [8-17]. The other basic steps are the already mentioned INFN proposal (1967) [2] and the publications from 1970 of the ADONE results, proving that the experimental set-up was working and that the acoplanar $(e^\pm \mu^\mp)$ method was not swamped by background [18-39].

2 — Lepton Pair Physics at CERN (1963-1968).

The 1967 INFN proposal [2] is based on a series of technological [4-10] and physical [11-17] research works performed at CERN where lepton pair production in hadronic processes was investigated.

The starting point was using collisions between hadrons (h),

$$h + h \rightarrow (e^+e^-) + X$$

but:

$$\frac{h + h \rightarrow (e^+e^-) + X}{h + h \rightarrow \text{ hadrons}} \approx 10^{-6}$$

thus a powerful rejection against hadrons, in order to detect (e^+e^-) pairs, was needed.

Comitato Nazionale per L'Energia Nucleare

ISTITUTO NAZIONALE DI FISICA NUCLEARE

Sezione di Bologna
67/1

INFN/AE-67/3
20 Marzo 1967

M. Bernardini, D. Bollini, E. Fiorentino, F. Mainardi, T. Massam, L. Monari, F. Palmonari and A. Zichichi (Bologna-Cern-Frascati collaboration) : A PROPOSAL TO SEARCH FOR LEPTONIC QUARKS AND HEAVY LEPTONS PRODUCED BY ADONE. -

Reparto Tipografico
dei Laboratori Nazionali di Frascati

Fig. 1.1:   Front page of ref. 2, i.e. the 1967 proposal to INFN by the Bologna-CERN-Frascati (BCF) group to search for a new Heavy Lepton, HL, at ADONE.

7.

By studying the most favourable mechanisms which could produce the heavy leptons we reach the following conclusion. If in the process

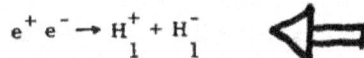

$$e^+ e^- \to H_1^+ + H_1^-$$

we set at an energy E such that the ratio

$$\frac{E}{M_{H_1}} \simeq 1.2$$

as can be seen from Fig. 6 the cross-section is around 10^{-32} cm^2. Moreover, the two produced H_1^+ and H_1^- are non relativistic and very slow in the laboratory-system, their $\gamma = E/M$ is in fact ~ 1.2. The most favoured decay channels, as far as we can say now, are probably

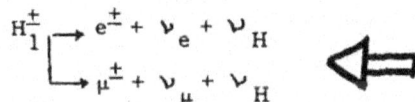

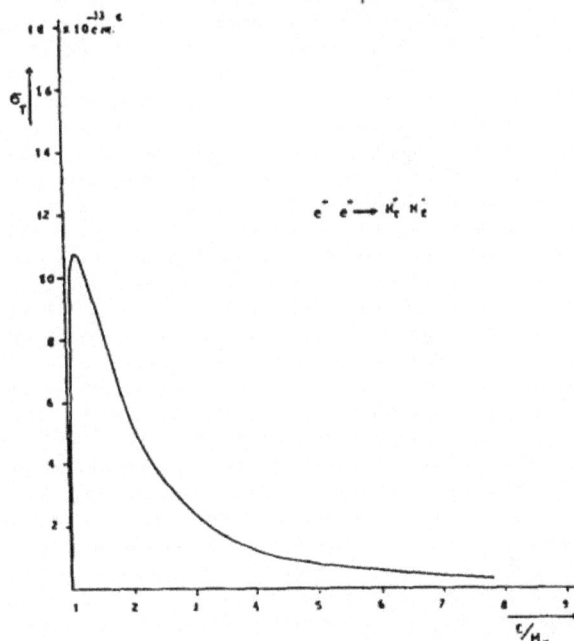

$$H_1^\pm \longrightarrow e^\pm + \nu_e + \nu_H$$
$$\longrightarrow \mu^\pm + \nu_\mu + \nu_H$$

FIG. 6 - Total cross-section for production of heavy leptons versus E/M_{H_1}.

Fig. 1.2: Showing the page of ref. 2 where the most favourable production process and decay channels foreseen to search for a new Heavy Lepton are reported.

CERN 63-25
Nuclear Physics Division
27th June 1963

ORGANISATION EUROPÉENNE POUR LA RECHERCHE NUCLÉAIRE
CERN EUROPEAN ORGANIZATION FOR NUCLEAR RESEARCH

A Telescope to Identify Electrons in the Presence of Pion Background

T. Massam, Th. Muller and A. Zichichi

G E N E V A

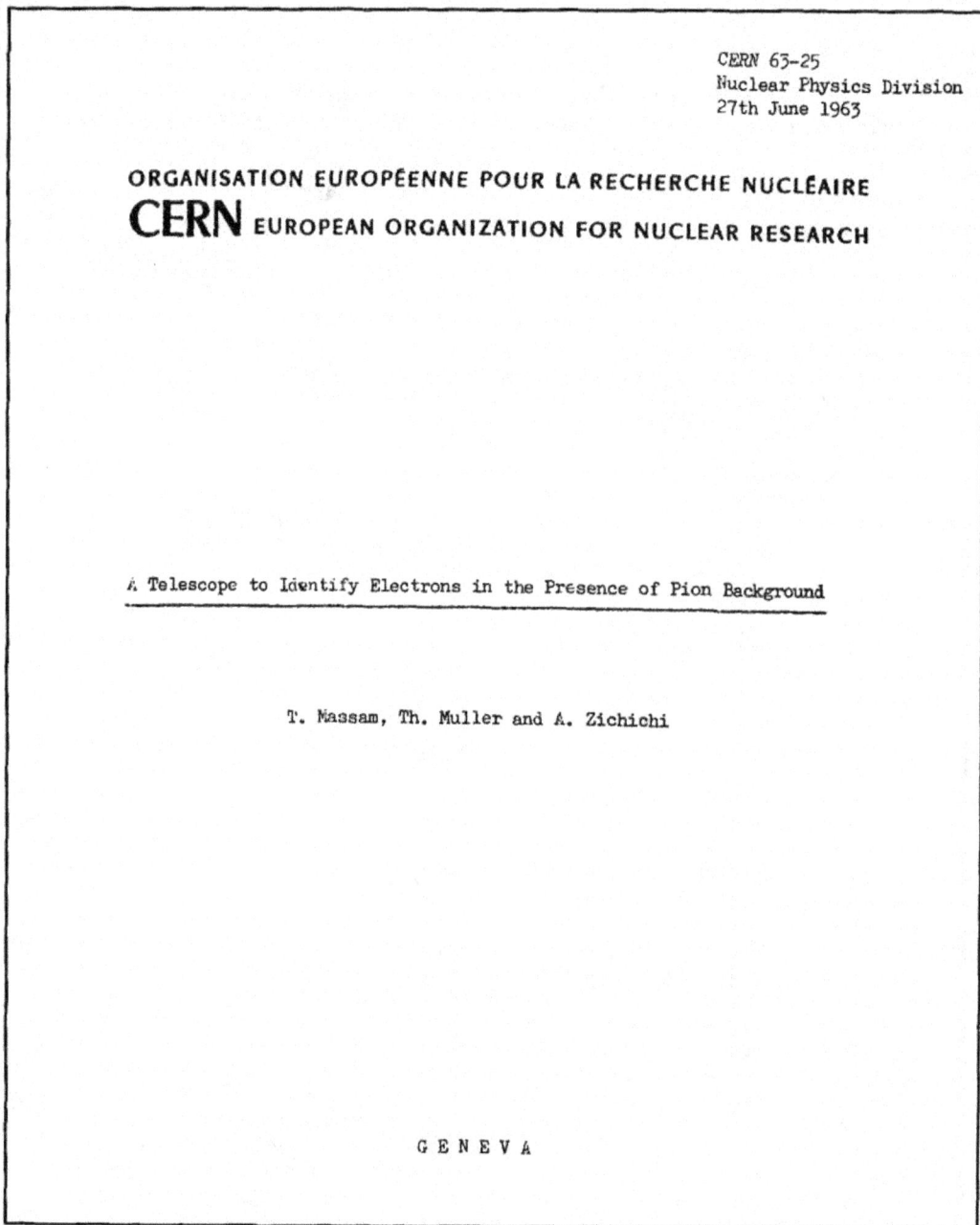

Fig. 2.1: Front page of ref. 4 where the innovative "early-shower development" (i.e. "pre-shower") method for π/e rejection was presented.

Figure 2.1 is the cover page of the CERN Yellow-Report [4] where the "pre-shower" technique was presented for the first time.

The "pre-shower" technique was able to reject pions in favour of electrons with a rejection power as good as

$$\pi/e \approx 10^{-3}.$$

Figure 2.2 shows the spectrum obtained for electrons and pions using this "pre-shower" technique associated with a lead-glass Čerenkov counter.

The Bologna-CERN group continued to improve the π/e rejection reaching the level of 10^{-4}. Our detector provided the most powerful π/e rejection and thus became the first example of an essential instrument now widely used in subnuclear physics: the calorimeter (electromagnetic and hadronic).

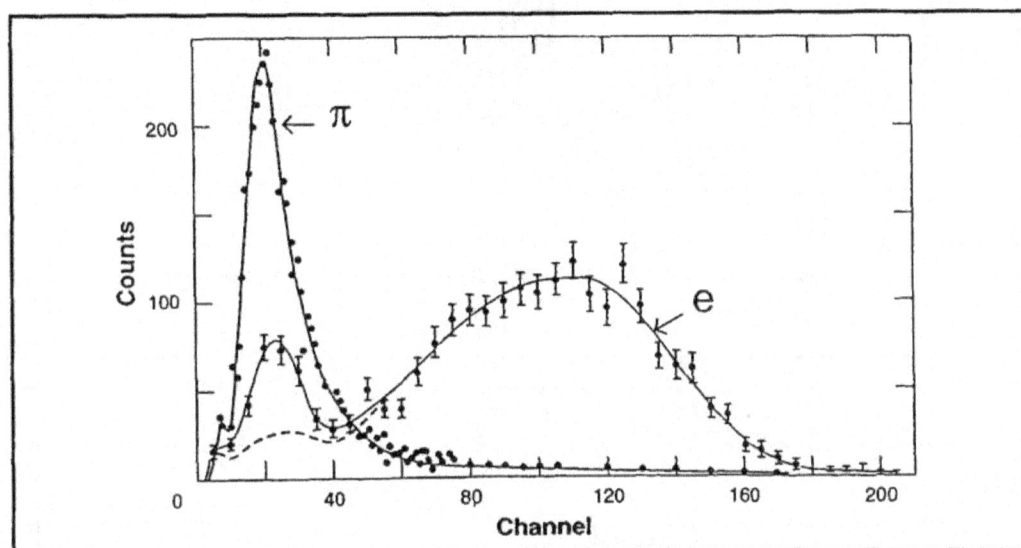

Fig. 2.2:   The preconverter pulse height spectra obtained in a test beam with the "pre-shower" technique, coupled to a total absorption lead-glass Čerenkov counter. This allowed one to reach a π/e rejection of 10^{-3} [4].

Figure 2.3 shows the new detector [5], designed and proven to work with a rejection power better than 10^{-3}. This electron detector consists of five elements, each one being made of a lead layer followed by a plastic scintillation counter and a two-gap spark chamber. The results are shown in Figure 2.4.

As can be seen from this figure, the rejection power of the new detector against pions is of the order of $4 \cdot 10^{-4}$ and the efficiency for electron detection varies from 75% to 85% (right scale). The energy resolution can be as good as 10%, in the energy range 1.1 GeV to 2.5 GeV.

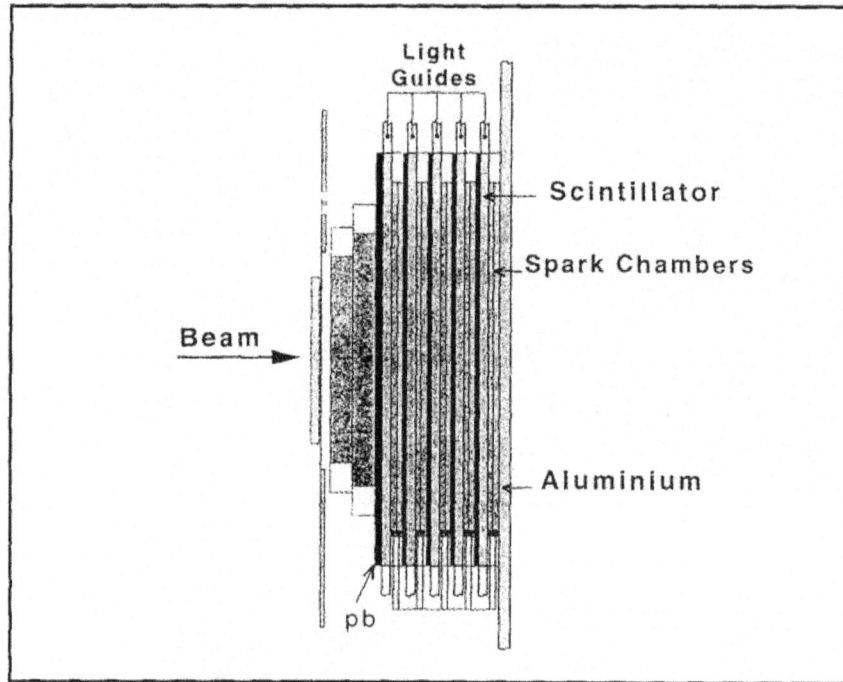

Fig. 2.3: The electron detector [5] consisting of five elements, each one being made of a lead layer followed by a plastic scintillation counter and a two-gap spark chamber.

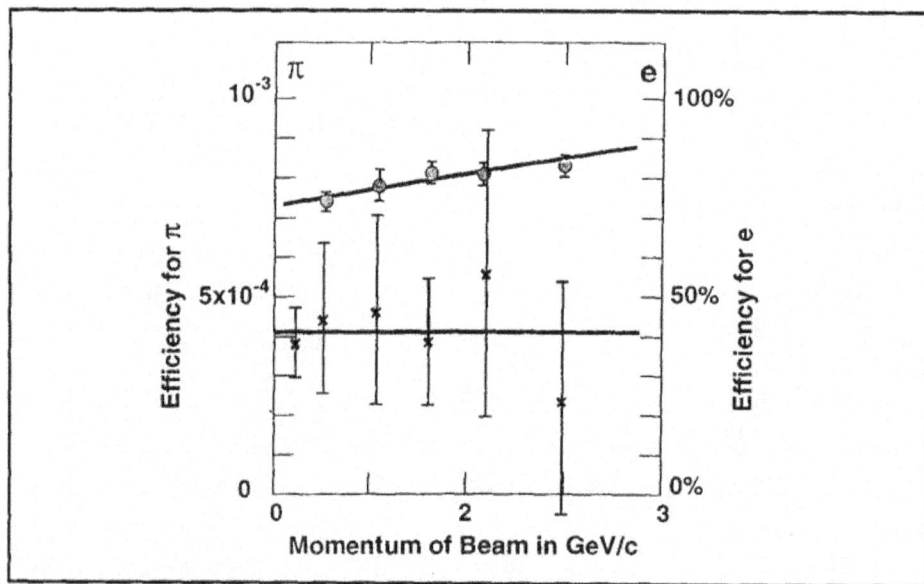

Fig. 2.4: The π (left scale) and e (right scale) efficiencies achieved with the set-up shown in Fig. 2.3 [5].

A crucial point of the research work at CERN was the simultaneous investigation, in addition to the (e^+e^-) pairs, of the $(\mu^+\mu^-)$ channel. As will be discussed below, these are experimental steps essential to the fundamental quest: the search for a new lepton carrying its own leptonic number.

As reported in our 1967 proposal to INFN, we were fully aware of the fact that, if a new lepton heavier than the muon (we had fixed its mass at the 1 GeV level) existed, this could have been produced in hadronic interactions, via time-like photons, but never seen because it would have decayed in a time interval as short as 10^{-11} sec. Notice that for this Heavy Lepton there could be no production process like $\pi \to \mu$, but only production via time-like photons. So, the crucial point was, first to study the abundance of time-like photons in hadronic interactions, and then to observe the typical signature of the Heavy Lepton decay, i.e. $(e\mu)$ pairs. The hadronic background had to be reduced relative to electrons and muons in order to allow the $(e\mu)$ "signature" to be detected. Figures 2.5 and 2.6 are the results of many years of work started in 1960 [4-17]. The first results were published in 1963 and referred to the (e^+e^-) channel in $(\bar{p}p)$ annihilation [11-12].

Fig. 2.5: General view of the experimental apparatus installed at the CERN PS to study the production of time-like photons yielding (e^+e^-), $(\mu^+\mu^-)$ and $(e^\pm\mu^\mp)$ pairs in $(\bar{p}p)$ annihilation [11-14].

Fig. 2.6:   General view of the experimental apparatus installed at the CERN PS to study the production of time-like photons in $(\pi^- p)$ interactions [15-17].

If in the time-like region the proton were point like, this reaction would have been a powerful source of time-like photons. The investigation of the proton form-factor in the time-like region was therefore important. Figure 2.7a shows the first page of our 1963 paper [11].

Notice that we used for this search the already described "pre-shower" technique coupled with a pair of total absorption lead-glass Čerenkov counters.

One year later we presented in Dubna our results with the $(\mu^+\mu^-)$ channel [13]. Figure 2.7b shows the first page of our contribution, with a sketch of the $(\mu^+\mu^-)$ set-up. The results with simultaneous detection of (e^+e^-) and $(\mu^+\mu^-)$ pairs were published a year later (1965) [14]. Figure 2.8 shows the front page of this paper.

The experimental investigation on the proton form-factor in the time-like region is a basic one in its own right; it supplements the information obtained by R. Hofstadter in the space-like region. For the Heavy Lepton search, it gave two important pieces of information: (i) the $(e\mu)$ pair technology worked as expected: in fact the enormous background had indeed been perfectly mastered; (ii) the primary production process for the time-like photon could not be $(\bar{p}p)$.

Volume 5, number 3   PHYSICS LETTERS   1 July 1963

SEARCH FOR TIME-LIKE STRUCTURE OF THE PROTON

M. CONVERSI *, T. MASSAM, Th. MULLER ** and A. ZICHICHI

CERN, Geneva

Received 5 June 1963

While the electromagnetic structure of the proton has been widely investigated for space-like four-momentum transfers through electron-proton scattering experiments, no information is available as yet on the proton structure in the time-like region. We report here the results of a search for the process:

$$\bar{p} + p \to e^- + e^+ , \qquad (1)$$

the rate of which depends on the unknown time-like form factors. The negative result of the experiment places an upper limit to the cross section for reaction (1), thus yielding evidence against a "point-like" proton in the time-like region.

The physical significance of an experimental study of process (1) and its connection with the intrinsic properties of the reacting particles have recently been pointed out by Zichichi et al. [1]. The cross section can be unambiguously predicted for a point-like proton. At an antiproton momentum of 2.5 GeV/c its value is ~ 0.25 μb.

The experiment was carried out at the CERN Proton Synchrotron with the apparatus *** sketched in fig. 1. The beam conditions and the geometry were chosen to maximise the rate of events from reaction (1). Consequently an antiproton momentum of 2.5 GeV/c with ± 1% momentum spread was chosen. The corresponding laboratory momenta for the secondary particles of a two-body process such as (1), were defined by the geometry of our apparatus to be in the range of 1.0 to 2.6 GeV/c.

Antiprotons from a partially separated beam **** enter (fig. 1) the cylindrical polyethylene target after crossing the monitoring plastic scintillators A and B and the anticoincidence counter D. D is a C_2H_4 gas Čerenkov counter sensitive to pions and lighter particles but not to $\bar{p}$ of 2.5 GeV/c momen-

* CERN Visiting Scientist, on leave of absence from the University of Rome, Italy.
** On leave from the Centre National de la Recherche Scientifique and Institut de Recherches Nucléaires, Strasbourg, France.
*** A more detailed description of the apparatus and of its performance will be given elsewhere.
**** This beam contained approximately equal numbers of $\bar{p}$ and π^- mesons. The $\bar{p}$ rate was about 10^3 $\bar{p}$/sec.

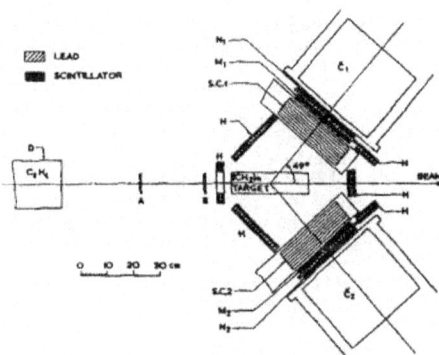

Fig. 1. Experimental layout. Plane view: D: 2 atm C_2H_4 Čerenkov counter; A, B: 1 cm thick, 8 cm diameter plastic scintillator, beam-defining counters; H: system of anticoincidence counters forming the "veto-house". They are made of two pieces of 1 cm plastic scintillator (seen by the same photomultipliers) with 1 cm lead in between; S.C.: 8 gap 0.025 mm Al foil spark chambers; M_1, M_2, N_1, N_2: 1 cm thick, 28 cm diameter plastic scintilator; $\check{C}_1$, $\check{C}_2$: lead-glass total absorption Čerenkov counters, 35 cm diameter, 30 cm thick (~ 8 radiation lengths).

tum. Incident $\bar{p}$ are defined by an anticoincidence pulse A$\bar{B}\bar{D}$.

The counters H consist of scintillator-Pb-scintillator sandwiches which form a "veto-house" around the target. The veto-house is sensitive to both charged particles and γ rays. It covers the whole solid angle not subtended by the two spark chambers, SC1, SC2, thus protecting against multi-pion events.

The essential part of the apparatus is a system of two similar telescopes. It is designed to accept the electron pair from process (1), while providing an efficient rejection of unwanted processes. Each telescope consists of three components:
i) an eight gap, thin foil spark chamber which allows kinematical reconstruction;
ii) counter M which is used to accept single

195

Fig. 2.7a: Front page of ref. 11 where the first evidence – using the e^+e^- channel – for the EM structure of the proton in the time-like region was presented.

Proceedings of the International Conference on "High-Energy Physics"
Dubna, USSR, 5–15 August 1964 (Atomizdat, Moscow, 1966), Vol. I, 857

PROTON ANTIPROTON ANNIHILATION INTO MUON PAIR

M. Conversi, T. Massam, Th. Muller, M. Schneegans and A. Zichichi
CERN. Geneva, Switzerland

(Presented by A. ZICHICHI)

The experiment I wish to talk about refers to the study of the process:

$$\bar{p} + p \longrightarrow \bar{\mu} + \mu.$$

As is well known the interest of this experiment is twofold:

i) determination of the electromagnetic structure of the proton in the time-like region;

ii) check of the equality of the electromagnetic properties of the muon and the electron for time-like momentum transfers.

The experiment has been performed at CERN, using the new high intensity separated $\bar{p}$-beam [1] whose characteristics are probably worth recording:

with 50% of the total circulating proton beam of the CERN Proton-Syncrotron (which has been during the last week 10^{12} protons/pulse) we get 50000 $\frac{\bar{p}}{burst}$ with a π-contamination: $\frac{\pi}{\bar{p}} = 3.5$; 250 m sec. spilling time and

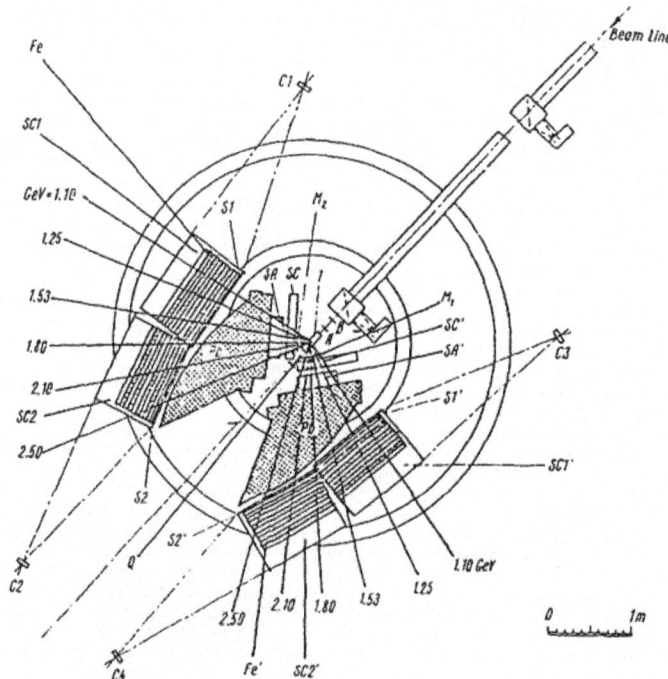

Fig. 1. Experimental set-up.

Fig. 2.7b: Front page of ref. 13 where the results on $\bar{p}p \rightarrow \mu^+\mu^-$ are presented.

M. CONVERSI, *et al.*
21 Novembre 1965
Il Nuovo Cimento
Serie X, Vol. 40, pag. 690-701

CERN
SERVICE D'INFORMATION
SCIENTIFIQUE

The Leptonic Annihilation Modes of the Proton-Antiproton System at 6.8 (GeV/c)² Timelike Four-Momentum Transfer.

M. CONVERSI (*), T. MASSAM, TH. MULLER (**) and A. ZICHICHI

CERN · Geneva

(ricevuto il 6 Novembre 1965)

This letter reports the investigation of the proton-antiproton annihilation processes:

(1) $$\bar{p} + p \to \bar{e} + e \, ,$$

(2) $$\bar{p} + p \to \bar{\mu} + \mu \, ,$$

which was suggested by ZICHICHI *et al.* (¹). The experiment was carried out at 2.5 GeV/c incident antiproton momentum, which corresponds to a timelike four-momentum transfer of 6.8 (GeV/c)². The choice of this q^2-value was dictated by the requirement of the maximum number of observable events predicted, using the knowledge of the $\bar{p}$ flux as a function of momentum at the CERN PS, together with the behaviour of the cross-section as a function of the $\bar{p}$ momentum for processes (1) and (2), which was calculated assuming that the proton-photon interaction is pointlike.

The aim of the experiment was twofold: firstly, to measure the electromagnetic form factors for timelike four-momentum transfers and secondly to check the equivalence of the electromagnetic interactions of the electron and the muon for timelike four-momentum transfers. In practice the cross-section was shown to be so low that only an upper limit for annihilation into lepton pairs of

(3) $$\sigma^{\text{point}}_{\bar{p}p \to l\bar{l}} < 0.54 \text{ nanobarn} \qquad (1 \text{ nanobarn} = 10^{-33} \text{ cm}^2)$$

could be obtained. This number should be compared with the values of the cross-section expected if the proton behaves as a point charge

(4) $$\sigma_{\bar{p}p \to l\bar{l}} = 242 \text{ nanobarn}$$

(*) Istituto di Fisica dell'Università, Roma.
(**) Institut de Recherches Nucléaires, Strasbourg.
(¹) A. ZICHICHI, S. M. BERMAN, N. CABIBBO and R. GATTO: *Nuovo Cimento*, 24, 170 (1962).

Fig. 2.8: Front page of ref. 14 on the leptonic (e^+e^-, $\mu^+\mu^-$) annihilation modes of ($\bar{p}p$) in the time-like region.

Our result established the non-point-like behavior of the proton in the time-like region. We proved that the cross section was 500 times lower than the expected point-like value. Thus, for practical purposes, there could not be ($e^{\pm}\mu^{\mp}$) pairs, not because there was no production of a Heavy Lepton carrying its own leptonic number, but because

$$\frac{\sigma\left[\overline{p}p \rightarrow \begin{cases} \rightarrow e^+e^- \\ \rightarrow \mu^+\mu^- \end{cases}\right]^{\text{time-like structure}}}{\sigma\left[\overline{p}p \rightarrow \begin{cases} \rightarrow e^+e^- \\ \rightarrow \mu^+\mu^- \end{cases}\right]^{\text{point-like}}} \leq \frac{1}{500}$$

Hence there were not enough time-like photons available for its detection. As pointed out by V.F. Weisskopf [40], the search for a new Heavy Lepton different from the known ones was outside the theoretical framework of that time, therefore a powerful set-up able to detect simultaneously electrons and muons had to be justified on the basis of accepted standard physics in order to obtain the needed financial support. For example the study of the electromagnetic (EM) structure of the proton in the time-like region was along this line [11-14]. The present status of this physics — 30 years later — is shown in Figure 2.9. Another example of "standard" physics is the $(\omega-\phi)$ mixing investigated using the (e^+e^-) channel alone, as reported in Figure 2.10 [15-17]. These are examples of the results obtained using the first large solid angle set-up able to detect simultaneously

$$\begin{cases} (e^+e^-) \\ (\mu^+\mu^-) \\ (e^\pm\mu^\mp) \end{cases}$$

pairs produced in hadronic processes.

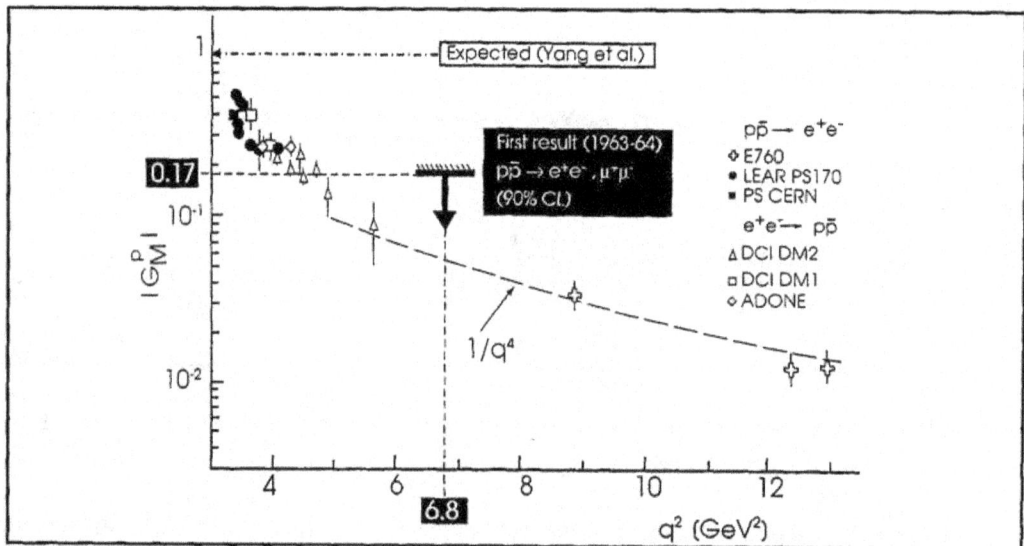

Fig. 2.9: The present status of the time-like EM form factor of the proton 30 years after the first evidence [11-14] by the Bologna-CERN group.

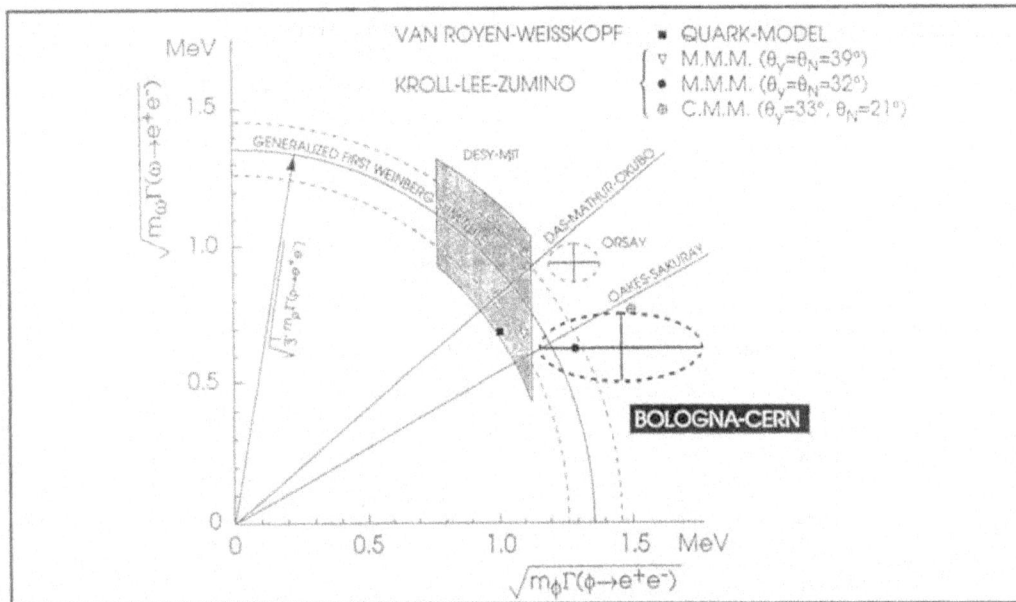

Fig. 2.10: The (ω-ϕ) mixing: the Bologna-CERN result [15-17] compared with the various theoretical expectations.

3 — Excited Leptons (e^*, μ^*) and Long Lived Heavy Muons (1960-1970).

To the best of our knowledge, during 1960-1970 nowhere but in my group, PAPLEP (at CERN) and BCF (at Frascati), was: 1) the idea of a Heavy Lepton with its own neutrino seriously studied, 2) the way to look for it investigated and worked out with a definite experimental proposal and, 3) proved to work.

The interest in Heavy Leptons, at that time, was very low. The theoretical trend was "there are too many leptons". Before the discovery $v_\mu \neq v_e$, Leon van Hove, at CERN, was saying: "Nature cannot be so stupid to have two neutrinos to do the same thing". When Lederman-Schwartz-Steinberger discovered that there were two neutrinos, the only Heavy Leptons considered worthy of some attention were of the type excited electrons (e^*) and/or muons (μ^*). None of these carries a new lepton number.

These excited states were expected to decay to the known leptons plus a γ:

$$e^* \rightarrow e\gamma$$
$$\mu^* \rightarrow \mu\gamma.$$

In order to recall the theoretical and experimental thinking of that time, let me show the first page of a theoretical paper [41] by Francis Low (Figure 3.1), where the heavy electrons and muons were theoretically investigated, and also the first page of an experimental paper [42] on a search for heavy stable muons by Martin Perl and collaborators (Figure 3.2) where heavy stable muons were searched for with a single-arm spectrometer (Figure 3.3).

VOLUME 14, NUMBER 7   PHYSICAL REVIEW LETTERS   15 FEBRUARY 1965

HEAVY ELECTRONS AND MUONS*

F. E. Low

Physics Department and Laboratory of Nuclear Science,
Massachusetts Institute of Technology, Cambridge, Massachusetts
(Received 15 January 1965)

The quantum electrodynamics (QED) of electrons, muons, and photons has so far been found to be in agreement with experiment.[1-3] This agreement has usually been expressed in terms of a fictitious "radius" down to which the theory has been found to hold.[4] In this language, an experimental deviation from the theory would reveal a "cutoff," or perhaps even a "cut-on."

A much more natural theoretical way of describing a breakdown of QED (and a more likely way for such a breakdown to occur) is in terms of coupling of electrons and muons to other particles.[5] This is consistent with the ideas of ordinary quantum field theory (or S-matrix theory), and is the only theoretically consistent way that we have to describe a real breakdown. In this language, continued experimental confirmation of the predictions of QED would be expressed in terms of upper limits to the coupling strengths and lower limits to the masses of hypothetical particles coupled to electrons, muons, and photons.

This point of view suggests a class of experiments which would search directly for such particles by looking for correlations in the mass spectrum of groups of final electrons and photons just as is done in strong-interaction physics. These experiments would be direct checks of QED. They would in many cases have the additional advantage of isolating the electrodynamic system from the nuclear target without the necessity of waiting for storage rings.

We discuss briefly three possible ways in which a breakdown might occur in the physics of electrons. Evidently, all remarks apply equally well to muons, although the experimental problems in that case are much harder.

(1) The electron might be coupled to a heavy electron, e', with a magnetic coupling of the form

$$\lambda \bar{\psi}_{e'} \sigma_{\mu\nu} \psi_{e} f_{\mu\nu} + \text{H.c.} \quad (1)$$

This is the most favorable case from the experimental point of view. Assuming a mass of the e' in the several hundred MeV range,

existing experiments are consistent with a coupling strength $\lambda \sim e/m_{e'}$, provided a reasonable cutoff is assumed and provided the decays $K^{\pm} \to e'^{\pm} + \nu$ and $K^0 \to e'^{\pm} + \nu + \pi^{\mp}$ are moderately forbidden. Otherwise, we must have $m_{e'} > 500$ MeV. The interaction (1) is neither minimal nor renormalizable. It would presumably be the low-energy manifestation of a minimal, renormalizable interaction (necessarily involving other particles) which would provide an automatic cutoff.

The simplest reaction to produce the e' would be

$$p + e \to p + e'$$
$$\quad \rightarrow e + \gamma \ (\tau \sim 10^{-21} \text{ sec}). \quad (2)$$

The e' would be observed as a sharp missing-mass peak in the recoil proton energy and angle distribution. This would be direct experimental evidence of an excited state of the electron. It could also be observed directly in a mass plot of the final $e + \gamma$.

The e' could also be produced by photons in the reaction

$$\gamma + p \to p + e + e'$$
$$\quad \rightarrow e + \gamma. \quad (3)$$

If the photons are tagged for energy, the e' could again be observed as a missing mass. With untagged photons, one could still observe a threshold in the missing mass as a function of maximum photon energy, or else detect directly a peak in the $e-\gamma$ mass spectrum. Depending on the precise experiment under consideration it might be advantageous to use a heavy target instead of hydrogen.

A further consequence of the existence of the e' (and of the minimal interactions coupling it to the electron) would be an anomalous Compton scattering of electrons and photons (at center-of-mass energies comparable to $m_{e'}$), as well as an anomalous electron-positron pair-production cross section at corresponding values of the electron-positron mass, possibly of the kind referred to in reference 3.

(2) The electron might be coupled to a boson,

Fig. 3.1:   Front page of ref. 41, where the Heavy Leptons considered are only excited electrons (e^*) or muons (μ^*).

PHYSICAL REVIEW   VOLUME 173, NUMBER 5   25 SEPTEMBER 1968

Search for New Particles Produced by High-Energy Photons*

A. Barna, J. Cox, F. Martin, M. L. Perl, T. H. Tan, W. T. Toner, and T. F. Zipf

Stanford Linear Accelerator Center, Stanford University, Stanford, California 94305

AND

E. H. Bellamy†

High Energy Physics Laboratory, Stanford University, Stanford, California 94305

(Received 5 April 1968)

A search for new particles which might be produced by photons of energy up to 18 GeV is described. No new particles were found. Calculations of the Bethe-Heitler process are described which make it possible to state that this experiment would have detected non-strongly-interacting particles whose mass and lifetime lay in a definite range, did they exist.

I. INTRODUCTION

WE have used the new Stanford linear electron accelerator to search for hitherto unknown elementary particles, particularly for particles which do not have strong interactions. The basic idea behind this search was that through the photoproduction of particle pairs, any charged particle can be created provided it has an antiparticle and that there is sufficient energy in the incident photon. The Stanford linear electron accelerator provides for the first time an intense source of high-energy photons—up to 18 GeV in this experiment. The experiment consisted of a momentum-analyzed secondary beam and a pair of differential gas Čerenkov counters which allowed particles of various masses in that beam to be detected. We were particularly interested in looking for non-strongly-interacting particles, and provision was made separately to detect strongly- and non-strongly-interacting particles.

In any search for new particles, the method of search limits in some ways the properties of the particles that might be found. This experiment was sensitive to charged particles with lifetimes greater than 5×10^{-8} sec, and with a production cross section at least 10^{-7} times that of the muon. Within these limitations, we have not found any new particles. We have made calculations, described in this paper, of the electromagnetic pair production of particles of arbitrary mass and zero spin. The results of these calculations and those of Tsai and Whitis[1] for spin-$\frac{1}{2}$ particles enable us to make the positive statement that if such non-strongly-interacting particles existed with a mass less than that of the proton and a lifetime similar to that of the kaon, we would have detected them.

II. GENERAL CONSIDERATIONS ON THE EXISTENCE OF ELEMENTARY PARTICLES

In our mind, there are two basic problems in elementary-particle physics. One is to understand and to calculate how the particles interact. The other is to learn what particles exist and to formulate rules which limit the possible kinds of particles. The two problems are related. This can be seen most clearly in the case of the strongly-interacting particles. The mesons and the numerous short-lived particles which appear as resonances in the strong interaction seem to be an intimate part of the interaction itself, so that one can expect that a correct theory of the interaction would also explain and predict the multitude of particles.

In the case of the particles which do not interact strongly, the situation is very different. The only known particles are the photon, the electron, the muon, and the two types of neutrinos. There is no understanding of why these particles and no others should exist, although the electromagnetic and weak interactions can be calculated. In particular, there is the puzzle of the existence of both the electron and the muon, particles so dissimilar in mass yet alike in all other aspects. Because the interactions can be calculated, it is possible to postulate the existence of a new particle and to calculate its lifetime and its effect on known processes as a function of its mass. Many authors have done this.[2] However, all such calculations make the basic assumption that no radically new feature enters into the interaction which could alter the result by orders of magnitude. As an example only, consider the effect of strangeness on the strong interaction. The muon-electron problem seems so little understood that some new concept as unlikely as strangeness was, may be required for its solution. We therefore believe that experimental searches for new particles should not be inhibited by preconceived ideas that short lifetimes are to be expected for massive, weakly-interacting particles. These ideas are based on our current understanding of the physics involved. This is true also of estimates of the production of hypothetical particles in specific processes. For example, the fact that K mesons are not observed to decay into heavy muons[3] means that according to

⬅

---

* Work supported by the U. S. Atomic Energy Commission.
† On leave from Westfield College, University of London, London, England.
[1] Y. S. Tsai and V. Whitis, SLAC Users Handbook, Part D (unpublished) and (private communication).

[1] F. E. Low, Phys. Rev. Letters 14, 238 (1965); F. J. M. Farley, Proc. Roy. Soc. (London) A285, 248 (1965); T. D. Lee and C. N. Yang, Phys. Rev. Letters 4, 307 (1960); J. Schwinger, Ann. Phys. (N. Y.) 2, 407 (1957); S. M. Berman et al., Nuovo Cimento 25, 685 (1962).
[2] A. M. Boyarski et al., Phys. Rev. 128, 2398 (1962); E. W. Beir, Ph.D. thesis, University of Illinois, 1966 (unpublished).

Fig. 3.2:   Front page of ref. 42 on the experimental search for heavy stable muons in photoproduction processes at Stanford.

173   PARTICLES PRODUCED BY HIGH-ENERGY PHOTONS   1397

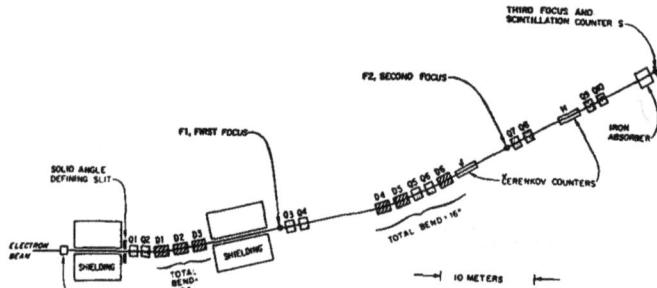

FIG. 6. Schematic diagram of the experiment.

a scintillation counter S was placed behind an iron absorber 5 ft thick. Weakly-interacting particles would have the signature HJS. The experiment consisted of fixing the beam momentum and varying the pressure of the gas in the Čerenkov counters in order to sweep through a range of masses, while recording HJ and HJS. The known particles provide indications of the operation of the system. In particular, the muons and pions provide a basic normalization of the experiment which does not depend upon the acceptance of the transport system or the efficiencies of the Čerenkov counters. Since the muon yield has been measured separately[14] and is understood theoretically, the muon normalization is particularly useful.

B. Apparatus

The target in which the secondaries were produced consisted of 3.6 radiation lengths of beryllium followed by ten radiation lengths of water-cooled copper, a further foot of beryllium, and ten radiation lengths of lead. The production of weakly-interacting particles in this target is adequately described by the calculations given in Sec. IV for production on beryllium, since there is very little particle production beyond the first 3.6 radiation length. The rest of the target was used to absorb the power (up to 20 kW) in the electron beam, and to reduce the number of electrons in the secondary beam to a few percent of the muon flux. Negatively charged secondaries from this target consist mainly of muons and pions. The composition of the beam at momenta of 5.0 and 9.0 GeV/c was measured to be approximately 70% muons, 30% pions.

The beam transport system shown in Fig. 6 was designed and built to provide a muon beam[15] for a muon-scattering experiment. It produces an almost dispersion-free beam in the Čerenkov counters with a diameter of less than 10 cm, a divergence of less than 4 mrad, and a momentum bite of ±1.5%. The second focus F2 is 212 ft from the target. Counter J was 19 ft

upstream from F2; counter H was 33 ft downstream from F2. The scintillation counter S was at the third focus, 63 ft downstream from F2.

The differential Čerenkov counters were modeled closely on a counter described by Kycia and Jenkins.[16] The present counters are designed to operate at pressures up to 960 psi. In this experiment, CO_2 was used at pressures up to 600 psi. Figure 7(a) is a schematic diagram of a counter. The radiator region is 80 in. long and the counter is designed to be used with beams up to 12.5 cm in diam. Čerenkov light from particles of the correct velocity is focused onto an annular ring aperture. The aperture is split in two across a diameter and the light from each half is collected separately onto two phototubes. A coincidence is required for a particle to be counted. The quartz windows are arranged so that a stray track in the general direction of the beam cannot go through both. Light which falls near, but not on, the annular aperture is reflected from a spherical mirror in which the aperture is set and is collected onto a phototube put in anticoincidence. Without this, a particle of the wrong velocity at an angle to the beam could be counted, as illustrated in Fig. 7(b). The width of the annular aperture was chosen to give an angular acceptance of ±10 mrad about a mean Čerenkov angle of 75 mrad. This dominated the mass resolution of the counters, giving $\Delta m/m \sim 0.075(p^2/m^2) \times 10^{-4}$, where m is the mass and p is the momentum of the particle. This resolution was adequate to separate out the peaks of the known particles, but allowed a finite mass range to be covered at each pressure setting and sufficient tolerance so that we did not have difficulty in operating the two counters together. The pressure vessels of the two counters were connected together by a common feed pipe. We found that no special precautions were necessary to make the mass peaks coincide in the two counters, although the counters were located out of doors and the ambient temperature varied from 5°C at night to 27°C during the day.

Block diagrams of the electronic circuits are shown in Fig. 8. The three tubes on each counter were fed through

[14] A. Barna *et al.*, Phys. Rev. Letters 18, 360 (1967).

[15] SLAC Users Handbook, Part D (unpublished); Stanford Linear Accelerator Center Laboratory Report No. SLAC-PUB 434 (unpublished); see also Ref. 13.

[16] T. F. Kycia and E. W. Jenkins, *Nuclear Electronics* (International Atomic Energy Agency, Vienna, 1963).

Fig. 3.3:   The page of ref. 42 showing the schematic diagram of the single-arm spectrometer used in the Stanford experiment.

Even later, when "theoretically wanted" Heavy Leptons came, in the attempt to avoid divergent neutrino cross sections, the Heavy Leptons advocated were again electron-like (E) and muon-like (M) [43].

Their difference from the known "electron" and "muon" was in the assignment of opposite electron and muon lepton numbers, respectively. The positively charged heavy electron (E^+) was given the positive lepton number identical to the lepton number of the known negative electron. Likewise, the positively charged heavy muon (M^+) was given the positive lepton number identical to the lepton number of the known negative muon. All this is shown in Figure 3.4.

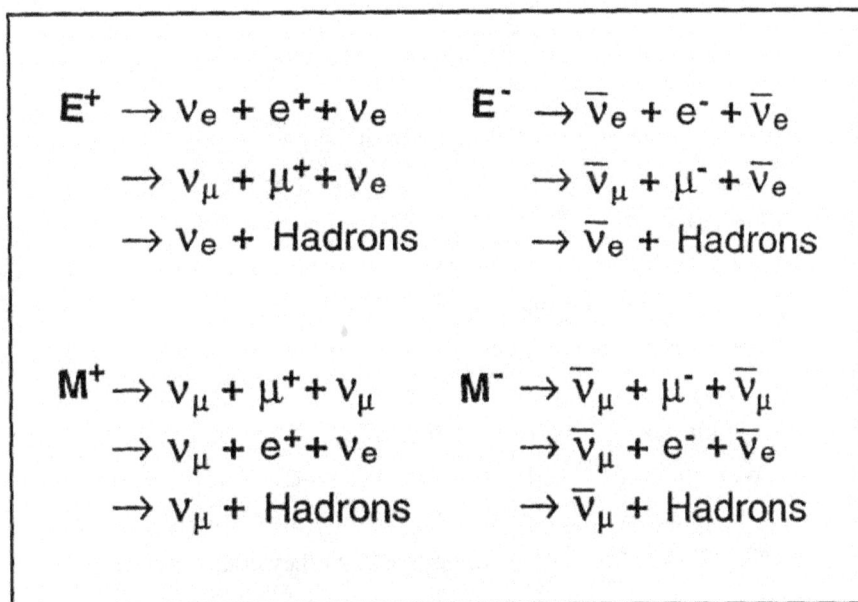

$$E^+ \rightarrow \nu_e + e^+ + \nu_e \qquad E^- \rightarrow \overline{\nu}_e + e^- + \overline{\nu}_e$$
$$\rightarrow \nu_\mu + \mu^+ + \nu_e \qquad \rightarrow \overline{\nu}_\mu + \mu^- + \overline{\nu}_e$$
$$\rightarrow \nu_e + \text{Hadrons} \qquad \rightarrow \overline{\nu}_e + \text{Hadrons}$$

$$M^+ \rightarrow \nu_\mu + \mu^+ + \nu_\mu \qquad M^- \rightarrow \overline{\nu}_\mu + \mu^- + \overline{\nu}_\mu$$
$$\rightarrow \nu_\mu + e^+ + \nu_e \qquad \rightarrow \overline{\nu}_\mu + e^- + \overline{\nu}_e$$
$$\rightarrow \nu_\mu + \text{Hadrons} \qquad \rightarrow \overline{\nu}_\mu + \text{Hadrons}$$

Fig. 3.4: The decay modes of the "theoretically wanted" electron-like (E) and muon-like (M) Heavy Leptons.

Heavy Leptons carrying their own lepton number, different from the electron and muon ones, were not within the theoretical framework of that time and, except for our search, also outside the experimental way of thinking.

4 — Design Considerations for our Proposal.

Suppose you had the idea of a new lepton (in addition to the known ones: e, μ) and suppose that this type of lepton is much heavier than the known ones — say 1 GeV — and carries its own leptonic number.

Question 1: What would be the best production process?

Question 2: How would it decay?

This type of lepton is supposed to be the least exotic particle. Therefore, it is like: a Standard Dirac Particle, insofar as its QED properties are concerned; and a Standard Fermi Particle, insofar as its weak properties are concerned. There is then nothing unknown in the production and decay rates, if its mass is given. The best production process would be via time-like photons:

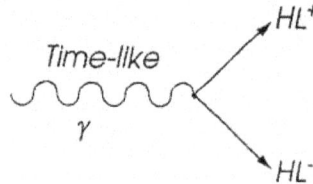

But a series of experiments carried out by us at CERN showed that "nucleons are very poor sources of time-like photons", as recalled again in a paper [Nuovo Cimento, 43, 227 (1966)] by T. Massam and myself.

Muons (the heaviest leptons known at that time) are very abundant in hadronic collisions. But a 1 GeV heavy new lepton carrying its own leptonic number would have escaped detection in all hadronic experiments. Let me quote what has been written in the 1967 INFN proposal [2]. "If Heavy Leptons exist would they have been detected?" "... The lifetime of a Heavy Lepton with 1 GeV mass would be of the order of 10^{-11} sec and could never have been detected as a decaying particle ..." "Moreover the production of μ is copious only because of the fact that it is the decay product of the π. There is no equivalent mechanism for the production of a 1 GeV Heavy Lepton: in proton-machines they could only be produced in pairs via time-like photons, a process of which the low rate has already been discussed. Moreover the lack of stability of this particle is consistent with its apparent absence." Therefore, production and decay of a new Heavy Lepton would be as follows.

i)   Production: if a Heavy Lepton exists, to search for it with hadron machines is hopeless. In fact, as we have seen in Section 2, it had been looked for at CERN where a powerful set-up for the detection of $(e^{\pm}\mu^{\mp})$ pairs was already working in 1964. The construction of this set-up started in the early sixties [40]. The production of time-like photons in hadronic collisions was found to be indeed very poor. The best way to produce Heavy Leptons was (and is) therefore via the reaction:

$$e^+e^- \to HL^+HL^-.$$

ii)   Decay: as the new Heavy Lepton is expected to be universally coupled to the known leptons, its decays will be:

$$HL^{\pm} \underbrace{\begin{array}{l} \longrightarrow e^{\pm} + \nu_e + \nu_{HL} \\ \longrightarrow \mu^{\pm} + \nu_{\mu} + \nu_{HL} \end{array}}$$

Many years of experimental, technological and phenomenological research work brought us to the conclusion that the only way to detect this new Heavy Lepton was via the acoplanar $(e^{\pm}\mu^{\mp})$ pairs produced in (e^+e^-) annihilations. But $e^+e^- \to e^{\pm}\mu^{\mp}\gamma$ will produce acoplanar $(e^{\pm}\mu^{\mp})$ pairs, so answers must be given to:

Question 1: Is $e^+e^- \to e^{\pm}\mu^{\mp}$ possible?

Question 2: Do radiative corrections really follow the "peaking approximation" (see Figure 4.1) ?

Question 3: Is QED correct at these high q^2-values?

Question 4: What is the background from $e^+e^- \to$ hadrons $\to e^{\pm}\mu^{\mp}X$?

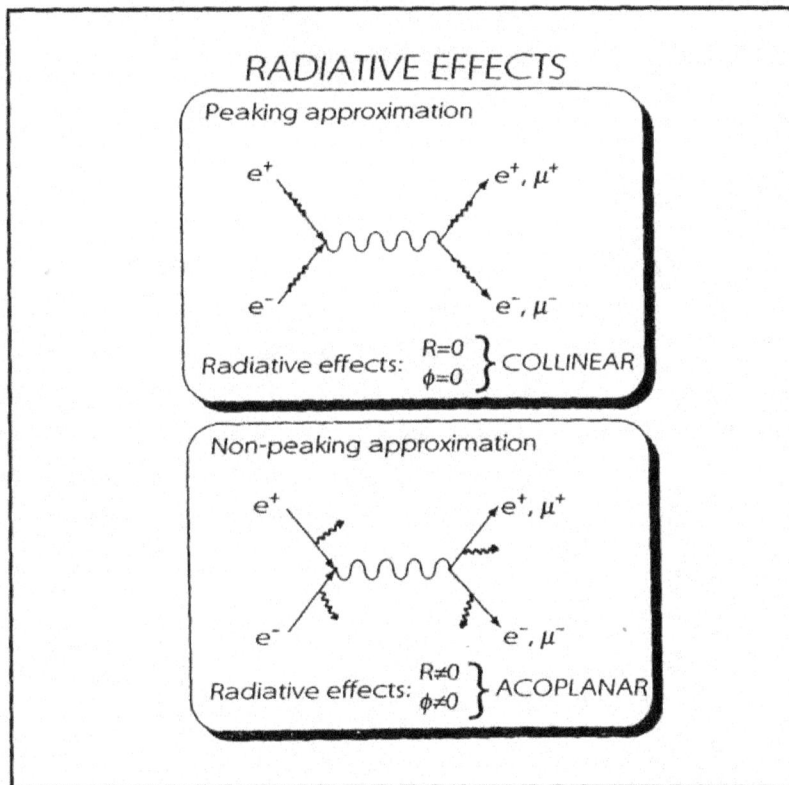

Fig. 4.1: Radiative effects with and without "peaking approximation" in (e^+e^-) annihilation [22]. R and ϕ are the acollinearity and acoplanarity angles, respectively. Obviously $\phi \neq 0$ implies $R \neq 0$, while the opposite is not true.

Moreover, in order to be sure that what is expected is indeed correctly observed in the experimental set-up, where acoplanar $(e^{\pm}\mu^{\mp})$ events have to be identified as a clear signature for the production of Heavy Lepton pairs, the proof is needed that the standard QED processes:

$$e^+e^- \rightarrow e^+e^-$$
$$e^+e^- \rightarrow \mu^+\mu^-$$

follow the theoretical QED predictions. This means a series of high-precision checks of QED in the same energy range. Therefore, the programme was very clear: the search for a new Heavy Lepton needed:

- the check of the validity of $e \neq \mu$ at high q^2-values;
- a detailed study of acoplanar radiative effects;
- high precision QED tests;
- the understanding of hadron production.

To detect $(e^{\pm}\mu^{\mp})$ pairs in an (e^+e^-) collider, a large solid angle detector is needed. This detector should be highly selective for electrons and muons, despite its large dimensions.

5 — The BCF Proposal (1967), the Experiment and the First Results (1970).

As already pointed out earlier, the preparatory work started at CERN after the first high precision measurements of the muon (g–2) [3] with the invention of the "pre-shower" technique [4-7], and the construction of the large solid angle set-up for detection of $(e^{\pm}\mu^{\mp})$ pairs in hadronic collisions where time-like photons able to produce Heavy Lepton pairs were searched for [11-14]. All this work culminated in 1967 with the INFN proposal by the Bologna-CERN-Frascati group [2] to search for Heavy Leptons at ADONE. In Section 1 we have shown the front page of our proposal (Figure 1.1) and the page where the production and decay reactions were given (Figure 1.2). The comprehensive programme to study:

$$e^+e^- \rightarrow e^+e^- \text{ or } \mu^+\mu^-$$
$$e^+e^- \rightarrow e^{\pm}\mu^{\mp}$$
$$e^+e^- \rightarrow h^+h^- + \text{anything}$$
$$e^+e^- \rightarrow HL^+ HL^- \rightarrow e^{\mp}\mu^{\pm} + \text{missing energy}$$

needed a large solid angle detector specially designed for this purpose. The INFN officials were asking for evidence that this type of large solid angle device could be built and could work as expected.

Our credentials were the CERN work on lepton-pair production in hadronic interactions.

On the basis of this work we could show that large solid angle detectors with high rejection power against different kinds of background had been built and proven to work in conditions worse than at Frascati. In fact, at CERN, a π-rejection at least at the level $\sim 10^{-3}$ was needed to allow the identification of electrons and muons:

$$\pi/e \leq 10^{-3}$$
$$\pi/\mu \leq 10^{-3}.$$

At Frascati the experimental conditions were expected to be much more favourable. Nevertheless, during the experiment (before the acoplanar (μe) method was shown to work), the scientific community in Frascati was divided. The majority was very much against both the experiment and the method. Most were saying that the search was going to be swamped by background: because in (e^+e^-) colliders there are plenty of electrons and spurious muons (from pion decays). The standard comment was: "Zichichi cerca farfalle", i.e. "Zichichi is looking for butterflies".

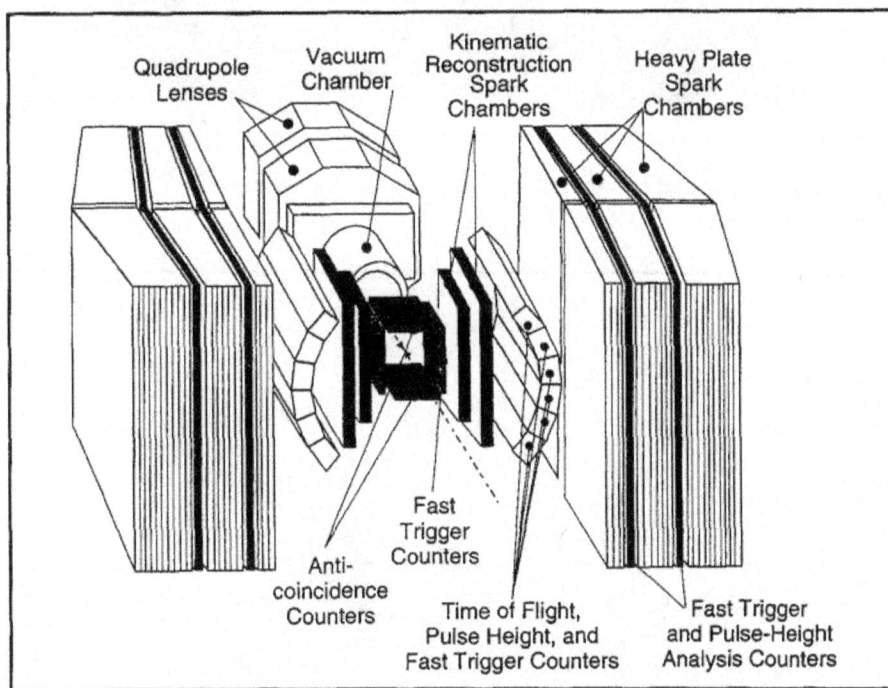

Fig. 5.1a: Perspective of the experimental apparatus used at ADONE [24], transverse to the (e^+e^-) beam-line.

The perspective of the experimental apparatus [24], transverse to the colliding (e^+e^-) beam-line, is shown in Figure 5.1a. The corresponding picture of the apparatus is shown in Figure 5.1b. The top-view [18] is shown in Figure 5.2a and its corresponding picture in Figure 5.2b.

Fig. 5.1b: A photograph of the apparatus whose detailed drawing is in Fig. 5.1a.

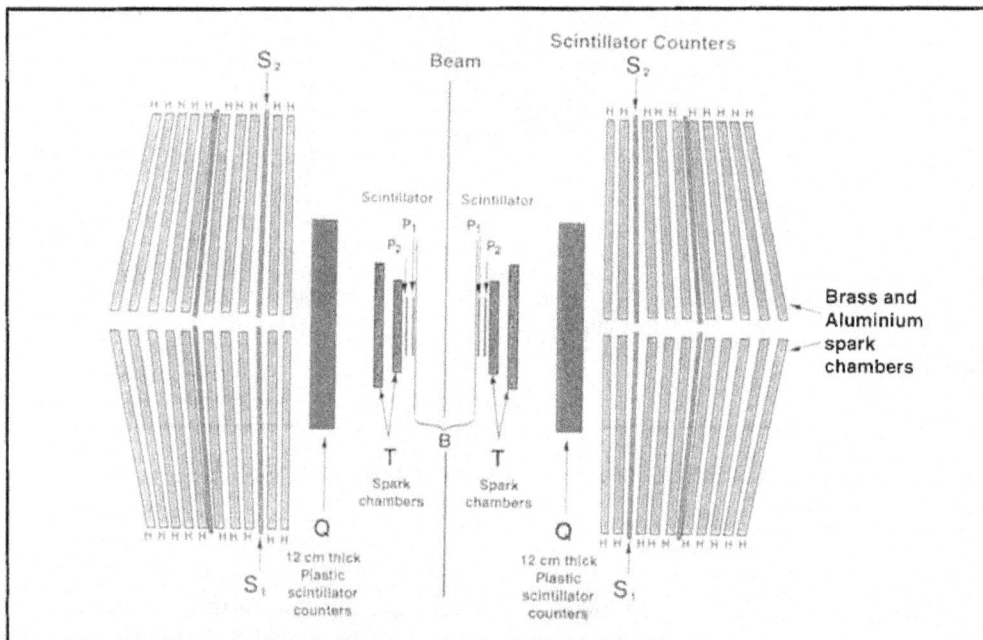

Fig. 5.2a: Top view of the ADONE apparatus [18].

Fig. 5.2b: The photograph corresponding to the Fig. 5.2a.

The essential elements of the set-up were:

    i)   a system of high precision thin-plate spark chambers for kinematic reconstruction;

    ii)   a system of high resolution time-of-flight counters (TOF);

    iii) a system of heavy-plate spark chambers.

Let me spend a few words on these essential elements.

The kinematic chambers provided a very clean peak at the vertex of the event thus rejecting a lot of unwanted background (Figure 5.3) [23].

The thick plastic scintillation counters (Q) provided high precision TOF measurements with $\Delta T = (\pm 0.35)$ nsec (Figure 5.4). Notice that the system consisted of 24 counters and the relative timing could be done within 0.1 nsec (Figure 5.5) [10].

An example of the value of this TOF system to reject cosmic muons is shown in Figure 5.6 [23]. Cosmic muons were at the 30% level, compared with the ADONE $(\mu^+\mu^-)$ rate. The TOF system was able to reject this background very efficiently.

The heavy-plate spark chambers allowed a clear distinction between "e" and "μ" [23]. A typical (e^+e^-)-pair event is shown in Figure 5.7. A typical $(\mu^+\mu^-)$-pair event is shown in Figure 5.8.

Notice how clean the electron-pair and muon-pair events are.

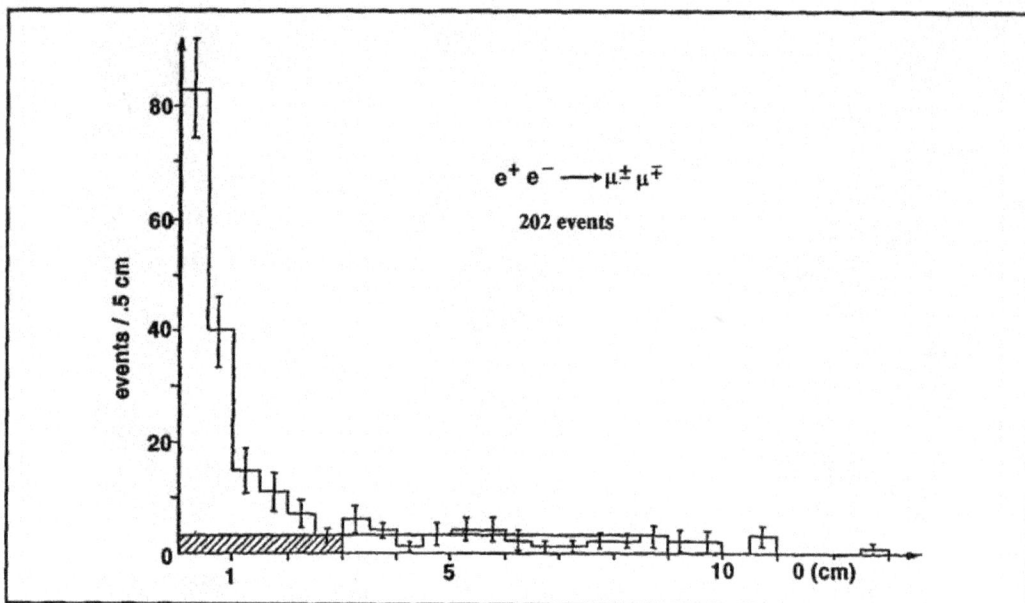

Fig. 5.3: Distribution of the minimum distance D between the beam axis and the reconstructed trajectory for $(\mu^{\pm}\mu^{\mp})$ pair events contaminated by cosmic-ray muons. The shaded area is our estimate of the cosmic-ray contribution in the selected ADONE events [23].

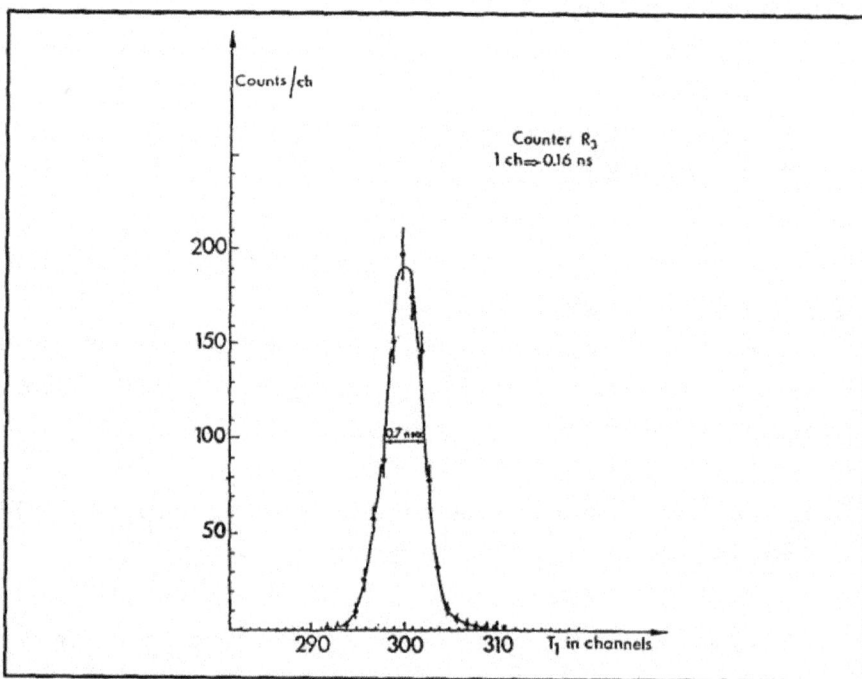

Fig. 5.4: Time-of-flight spectrum obtained using a relativistic pion beam, showing the achieved resolution [10]: $\Delta T = \pm 0.35$ nsec.

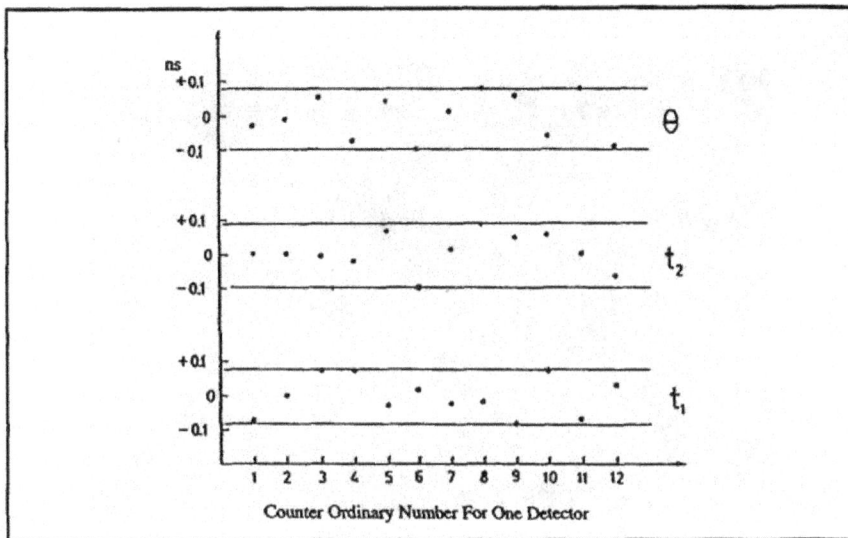

Fig. 5.5: Measured time difference showing the accuracy ($\pm$ 0.1 nsec.) of the relative timing amongst the various TOF counters [10].

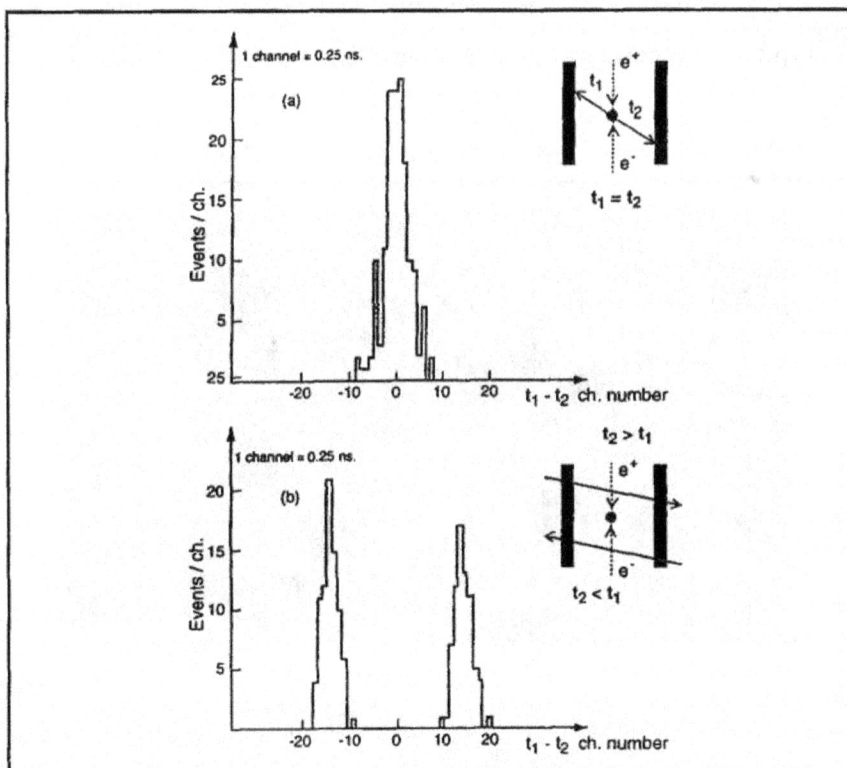

Fig. 5.6: Time-of-flight distribution for: a) selected ADONE events; b) cosmic rays with ADONE off [23].

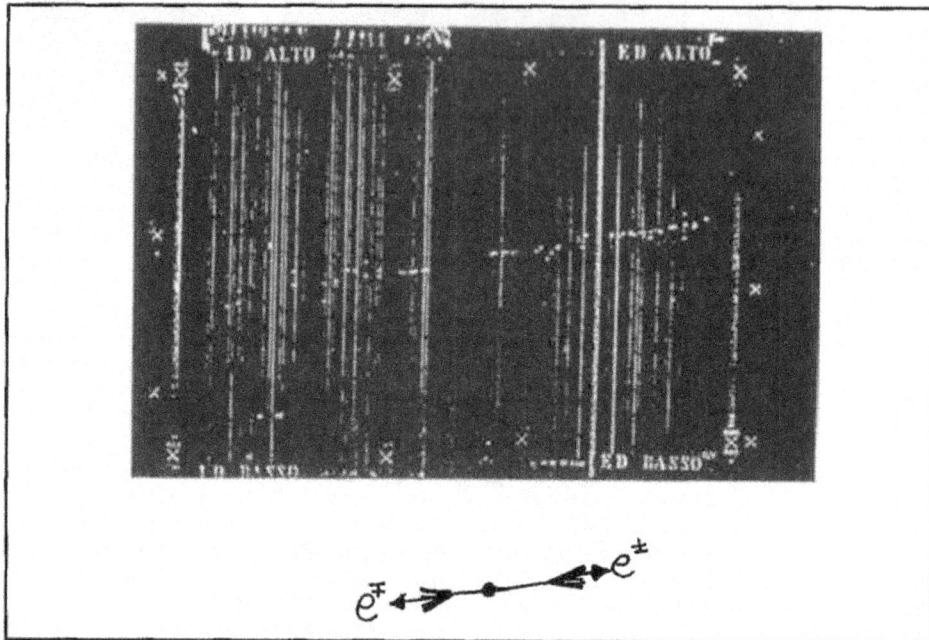

Fig. 5.7: Typical electron-pair event, as observed in the set-up of Figs. 5.1 and 5.2 at ADONE.

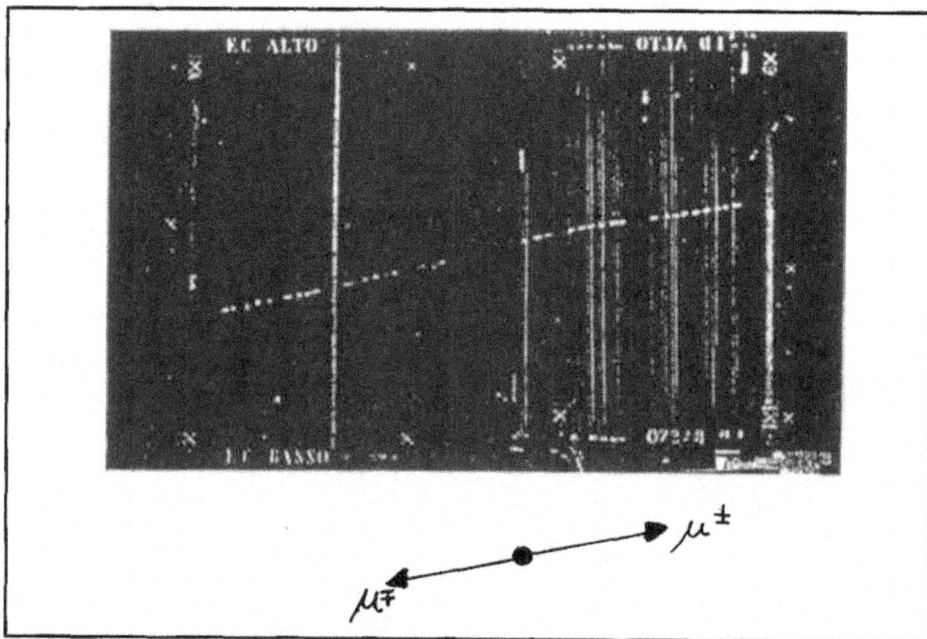

Fig. 5.8: Typical μ-pair event, as observed in the set-up of Figs. 5.1 and 5.2 at ADONE.

This was the status of our research through 1970, when the first results were published by the BCF group:

i)   on the Heavy Lepton searches at ADONE [19];

ii) on a series of experiments to demonstrate that, if $(e^{\pm}\mu^{\mp})$ acoplanar events were found, they could not be due to the violation of the leptonic number $(e \neq \mu)$ at high q^2-values (time- and space-like), coupled to acoplanar radiative corrections. These detailed studies showed that everything was following QED expectations for (e^+e^-) and $(\mu^+\mu^-)$ pairs produced in (e^+e^-) interactions [18, 20].

In Figure 5.9 the front page of the paper [19] on the first limit for Heavy Lepton production is shown, and in Figure 5.10a, the one relative to the first paper on the study of the validity of the leptonic number selection rules $(e \neq \mu)$ [18]. Figure 5.10b shows the relevant page of this paper, where the quantitative result is given.

Let me report what has been published by us on the test of lepton-number conservation from $e^+e^- \rightarrow e^{\pm}\mu^{\mp}$ [20]. "The study of this reaction allows one to establish the validity of the leptonic number selection rules for high space-like and time-like q^2-values. Collinear events with an electron and a muon in the final state would represent a proof that the presently known leptonic selection rules are violated.

No events of type $e^+e^- \rightarrow e\mu$ were found, and from the total number of observed lepton pairs, we get:

$$\frac{e^+e^- \rightarrow e^{\pm}\mu^{\mp}}{e^+e^- \rightarrow \text{lepton + antilepton}} \leq 2 \cdot 10^{-3} \quad \text{with 95\% confidence ...}$$

Background sources for this reaction are proved to be absent, from beam-gas interaction, cosmic radiation, or from simulation by $(e^{\pm}e^{\mp})$ or $(\mu^{\pm}\mu^{\mp})$ final states."

In the same year 1970 other results were reported by the BCF group at the Erice School. The front page of the corresponding preprint is reproduced in Figure 5.11.

These results were later published in [20-39]: Physics Letters, Nuovo Cimento and the Proceedings of the 1970 "Ettore Majorana" International School of Subnuclear Physics (issued in 1971). Figure 5.12 is the first page of the 1970 Erice Lecture, as appeared in the Proceedings one year later [20].

The results proved that, in an (e^+e^-) collider, the acoplanar $(e^{\pm}\mu^{\mp})$ method to search for a new Heavy Lepton accompanied by its own neutrino was clean and not swamped by background. From the first limit on *HL* published in 1970 [19] the notion of a Heavy Lepton (indicated first as $H_l^{\pm}$ later as $HL^{\pm}$, and finally as $L^{\pm}$ in the publications by the BCF group) and the method to look for it, i.e. the detection of acoplanar $(e^{\pm}\mu^{\mp})$ pairs, became known the world over and stimulated other searches, at Frascati and elsewhere.

LETTERE AL NUOVO CIMENTO   VOL. IV. N. 24   12 Dicembre 1970

Limits on the Electromagnetic Production of Heavy Leptons.

V. ALLES-BORELLI, M. BERNARDINI, D. BOLLINI, P. L. BRUNINI,
T. MASSAM, L. MONARI, F. PALMONARI and A. ZICHICHI

CERN - Geneva
Istituto Nazionale di Fisica Nucleare - Sezione di Bologna
Istituto di Fisica dell'Università - Bologna
Laboratori Nazionali del CNEN - Frascati (Roma)

(ricevuto il 6 Novembre 1970)

A comparison between the long list of hadronic states and the very short list of leptonic states exposes one of the most striking puzzles of particle physics. It is therefore in order to ask whether heavy leptons could have been detected in previous experiments. If universality for the coupling of this new lepton to the known leptons is assumed, then the lifetime of a heavy lepton is predicted to be $\sim 3 \cdot 10^{-10}$ s at 500 MeV and $\sim 2 \cdot 10^{-11}$ s at 1000 MeV mass values. Thus, for masses in the region of 1 GeV, they could never have been detected as a decaying quasi-stable particle, but only as a resonance in the lepton system. Furthermore, it should be noted that the production of the heaviest lepton known so far (the muon) is copious only because it is the decay product of a very commonly produced particle, the π. There is no equivalent mechanism for the production of a heavy lepton with a mass of about 1 GeV. In proton machines they could only be produced in pairs via timelike photons, but it is known that nucleons are very poor sources of timelike photons [1], owing to the rapid decrease of their form factors as the four-momentum transfer increases [2].

The most favourable mechanism for the production of a heavy lepton HL is

(1) $$e^+ e^- \rightarrow HL + \overline{HL} ,$$

which, in the one-photon approximation, is described by the Feynman diagram

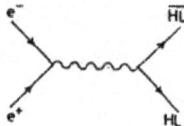

---

[1] T. MASSAM and A. ZICHICHI: *Nuovo Cimento*, **44 A**, 309 (1966). The deep inelastic effect discovered at SLAC could alter this statement. However, as yet no firm experimental results exist on this possible consequence of the SLAC results. This point will be discussed further in a forthcoming note.
[2] M. CONVERSI, T. MASSAM, TH. MULLER and A. ZICHICHI: *Nuovo Cimento*, **40 A**, 690 (1965).

Fig. 5.9:   Front page of ref. 19 on the first limit for the production of a new Heavy Lepton, HL, in $(e^+ e^-)$ annihilation.

V. Alles-Borelli, *et al.*
12 Dicembre 1970
Lettere al Nuovo Cimento
Serie I, Vol. 4, pag. 1151-1155

Validity of the Leptonic Selection Rules
for the (μeγ) Vertex at High Four-Momentum Transfers.

V. Alles-Borelli, M. Bernardini, D. Bollini, P. L. Brunini,
T. Massam, L. Monari, F. Palmonari and A. Zichichi

CERN - Geneva
Istituto Nazionale di Fisica Nucleare - Sezione di Bologna
Istituto di Fisica dell'Università - Bologna
Laboratori Nazionali del CNEN - Frascati (Roma)

(ricevuto il 6 Novembre 1970)

Using the Frascati (e^+e^-) colliding-beam machine (ADONE) [1] we have performed an experiment to look for the possible existence of the process

$$e^+e^- \to \mu^\mp e^\pm , \tag{1}$$

which, in the one-photon approximation, can take place if at the μ-e-γ vertex the currently known leptonic selection rules are violated. The available experimental information does not allow a distinction to be made between the two alternative classes of selection rules [2] which distinguish the « electron world » from the « muon world »; namely: *a*) two additive selection rules; *b*) an additive and a multiplicative selection rule. Both sets of rules would be violated by the existence of process (1). For very low q^2 values ($q^2 \simeq 0.01$ (GeV)^2) it is known that process (1) is strongly depressed. Examples are the unobserved processes

$$\mu^\pm \to e^\pm + \gamma , \tag{2}$$

$$\mu^- + \text{nucleus} \to (\text{nucleus})' + e^- . \tag{3}$$

However, as nobody knows the reason for the existence of these two leptonic quantum numbers, it is of interest to study their validity *vs.* q^2.

The data presented here result from the analysis of e^+e^- collisions, with energies ranging from 0.8 GeV to 1 GeV, in the angular range $(45 \div 135)°$. The corresponding range of momentum transfer associated with the photon at the (μeγ) vertex is $((0.38 \div 3.4)$ (GeV)^2) spacelike, and $((2.6 \div 4.0)$ (GeV)^2) timelike.

---

[1] F. Amman *et al.*: *Notiziario del CNEN*, **10**, 16 (March 1964); *ADONE, the Frascati 1.5 GeV electron-positron storage ring*, LNF-65/26 (Frascati, 30/8/1965).
[2] A. Zichichi: *Suppl. Nuovo Cimento*, **3**, 894 (1965).

Fig. 5.10a: Front page of ref. 18 on the study of the validity of the leptonic number selection rules $(e \neq \mu)$.

VALIDITY OF THE LEPTONIC SELECTION RULES ETC.   1155

The selection of e's and μ's turns out to be simple in this set-up because they carry strikingly different and clearly distinguishable signatures. Electrons and positrons give very clear electromagnetic showers in the heavy-plate chambers. A typical electron-pair event is shown in Fig. 2 and a typical muon-pair event is shown in Fig. 3.

The analysis of the data, taken at the various energies and with the luminosities quoted in Table I, has presented no good candidate for a (muon-electron) pair event.

From the total number of observed leptons pairs, $N(\text{lepton}+\text{antilepton}) = 1762 \pm 45$, we obtain the following upper limit for the branching ratio

(4)
$$\frac{e^+e^- \to e^\pm\mu^\mp}{e^+e^- \to \text{lepton} + \overline{\text{lepton}}} < 2\cdot 10^{-3}$$

with 95% confidence.

If we express in terms of a form factor $F^{\mu e\gamma}(q^2)$ a possible violation of the leptonic selection rules, our results imply that

$$\left| \underbrace{\int_{0.38\,(\text{GeV})^2}^{2.4\,(\text{GeV})^2} F^{\mu e\gamma}(q^2)\,dq^2}_{\text{spacelike}} + \underbrace{\int_{2.56\,(\text{GeV})^2}^{4.0\,(\text{GeV})^2} F^{\mu e\gamma}(q^2)\,dq^2}_{\text{timelike}} \right|^2 < 2\cdot 10^{-3}\,.$$

If we assume that the leptonic selection rules are violated only in the spacelike range then

$$\left| \int_{0.38\,(\text{GeV})^2}^{2.4\,(\text{GeV})^2} F^{\mu e\gamma}(q^2)\,dq^2 \right|^2 < 2\cdot 10^{-3}\,;$$

while if only the timelike range is responsible for the violation, the limit turns out to be

$$\left| \int_{2.56\,(\text{GeV})^2}^{4.0\,(\text{GeV})^2} F^{\mu e\gamma}(q^2)\,dq^2 \right|^2 < 10^{-2}\,.$$

* * *

This experiment is a collaboration between the University of Bologna and CERN, started at the time when Prof. V. F. WEISSKOPF was Director General at CERN and Prof. G. PUPPI Director of Research. It is thanks to their support that this collaboration has been established. We are very grateful to them and would like to thank Profs. B. P. GREGORY, P. PREISWERK and G. COCCONI for having maintained the support.

The technicians of our group have devoted their greatest efforts towards the success of the experimental programme, and we would like to express our appreciation for the skilful work and friendly assistance at all stages of the experiment to G. BACCHERINI, J. BERBIERS, F. MARTELLI, F. MASSERA, G. MOLINARI and O. POLGROSSI.

Finally our gratitude is extended to those who are responsible for the invention, design, construction and running of the colliding-beam facility: Profs. B. TOUSCHEK, F. AMMAN, M. PLACIDI, and their collaborators.

Fig. 5.10b: The page of ref. 18 where the relevant quantitative result is given, i.e. the 95% CL limit on the $(e^+e^- \to e^\pm\mu^\mp / e^+e^- \to l\,\bar{l})$ branching ratio, with $l = e, \mu$.

EUROPEAN ORGANIZATION FOR NUCLEAR RESEARCH

STUDY OF CHARGED FINAL STATES PRODUCED IN e^+e^- INTERACTIONS

V. Alles-Borelli, M. Bernardini, D. Bollini,

P.L. Brunini, E. Fiorentino, P.L. Frabetti,

T. Massam, L. Monari, F. Palmonari and A. Zichichi

(Presented by A. Zichichi)

CERN, Geneva, Switzerland

Istituto Nazionale di Fisica Nucleare, Bologna, Italy

Istituto di Fisica dell'Università, Bologna, Italy

Laboratori Nazionali di Frascati, Italy

Paper presented at the E. Majorana International School of Physics

Erice 1 - 19 July 1970

To be published in: "Elementary Processes at High Energy" - Academic Press -

New York and London

Fig. 5.11: Front page of the CERN preprint (later published as ref. 20) where all the relevant results by the BCF group on lepton and hadron production in (e^+e^-) annihilation at ADONE were presented.

ELEMENTARY PROCESSES
AT HIGH ENERGY

"Ettore Majorana" International Centre for Scientific Culture
1970 International School of Subnuclear Physics
a NATO - MPI - MRST Advanced Study Institute
Sponsored by the Regional Sicilian Government
and the Weizmann Institute of Science
Erice, July 1-19

Study of Charged Final States
Produced in e^+e^- Interactions

V. ALLES BORELLI, M. BERNARDINI, D. BOLLINI, P. L. BRUNINI,
E. FIORENTINO, T. MASSAM, L. MONARI, F. PALMONARI and A. ZICHICHI (*)

CERN - Geneva
Istituto Nazionale di Fisica Nucleare - Bologna
Istituto di Fisica dell'Università - Bologna
Laboratori Nazionali di Frascati - Frascati (Roma)

Using the Frascati colliding beam facility, ADONE, the following reactions have been studied:

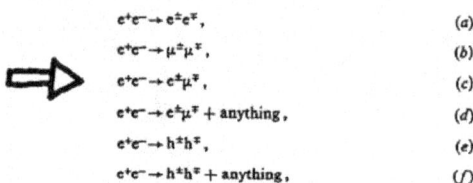

$$e^+e^- \to e^\pm e^\mp, \qquad (a)$$
$$e^+e^- \to \mu^\pm \mu^\mp, \qquad (b)$$
$$e^+e^- \to e^\pm \mu^\mp, \qquad (c)$$
$$e^+e^- \to e^\pm \mu^\mp + \text{anything}, \qquad (d)$$
$$e^+e^- \to h^\pm h^\mp, \qquad (e)$$
$$e^+e^- \to h^\pm h^\mp + \text{anything}, \qquad (f)$$

where h stands for « hadron ».

1. The experimental set-up.

Figure 1 shows a simplified sketch of the experimental set-up, which consists of four similar telescopes, two on each side of the colliding beam

(*) Presented by A. Zichichi.

1971

AP

ACADEMIC PRESS NEW YORK AND LONDON

Fig. 5.12: Front page of ref. 20 (corresponding to the CERN preprint already issued in 1970).

6 — The Heavy Lepton and the Acoplanar $(e\mu)$ Method from 3 GeV to Higher Energy (1971-1975).

Our 1970 paper [19] triggered a lot of interest. For instance Martin Perl published a paper in Physics Today [44] (July 1971), where he refers to this first Heavy Lepton paper (1970). The relevant page where he quotes Reference 5 (our 1970 results, i.e. Reference 19 of the present report) is reproduced in Figure 6.1.

for this second direction to be fruitful, one must either measure known properties with greater precision or one must measure properties that have not been previously measured. The recent high-precision measurements[1] of the gyromagnetic ratio of the muon are an illustration of the first type of measurement. The deep inelastic-scattering experiment (see figure 1), which I will describe later, is an illustration of the second type of measurement.

A larger family?

To the question: "Are the muon and the electron part of a larger family of charged leptons?" we must give an unsatisfactory answer; as far as we know there are no other charged leptons, but this knowledge does not go very far. The evidence can be summarized:

▶ Numerous experiments, many having to do with the decay of the K meson, have shown *no* additional leptons with masses below 0.5 GeV.

▶ No leptons with masses above 0.5 GeV have been found. But all searches for such particles have been incomplete. One reason for this incompleteness is that as the mass of the charged lepton increases, its lifetime becomes shorter and shorter, making direct detection more and more difficult.[4] Thus a lepton with a mass near 1 GeV will have a lifetime of about 10^{-11} seconds due to the decay processes in equations 2 and 3. A second reason for the incompleteness of past searches is that reactions that could copiously produce heavy charged leptons were not available.

Fortunately these problems can be overcome in the newly developed electron–positron colliding-beam accelerators where charged leptons can be copiously produced through the process[5]

$$e^+ + e^- \rightarrow \mu'^+ + \mu'^-$$

Within five years, through this process, we shall know if the electron–muon family has additional members with masses in the several-GeV ranges.

Static properties

One of the beautiful aspects of the search for muon–electron differences is the tremendous range of techniques that have been used. These techniques range from radiofrequency measurements of the hyperfine structure of muonium[2,3] (4500 MHz equivalent to 1.9×10^{-14} GeV) to measurements of muon-proton elastic and inelastic scattering at energies above 10 GeV. In surveying the results from this range of techniques I will first discuss the static properties of the muon and then discuss

Measuring muon–proton inelastic scattering at SLAC. Muons scattered in a hydrogen target (out of sight at the left) pass through the magnet (left center) and then through six spark chambers (right). Mirrors provide the stereoscopic view (lower set of tracks). See also cover photograph and figure 6. Figure 1

PHYSICS TODAY / JULY 1971   35

Fig. 6.1: The page of the article by M. Perl [44] where the first Heavy Lepton paper published by the BCF group [19] is referenced as ref. 5 (superscript).

CERN
SERVICE D'INFORMATION
SCIENTIFIQUE

Volume 36B, number 2   PHYSICS LETTERS   23 August 1971

EXPERIMENTAL PROOF OF THE INADEQUACY OF THE PEAKING APPROXIMATION IN RADIATIVE CORRECTIONS

V. ALLES BORELLI, M. BERNARDINI, D. BOLLINI,
P. L. BRUNINI, E. FIORENTINO, T. MASSAM, L. MONARI,
F. PALMONARI and A. ZICHICHI
CERN, Geneva, Switzerland
Istituto Nazionale di Fisica Nucleare, Bologna, Italy
Istituto di Fisica dell'Università, Bologna, Italy
Laboratori Nazionali di Frascati, Italy

Received 5 July 1971

49 e^+e^- non-collinear, non-coplanar events have been observed in a study of 1824 e^+e^- interactions at total centre-of-mass energies from 1.6 GeV to 2.0 GeV. The inadequacy of the peaking approximation in radiative corrections is measured to be $(2.8 \pm 0.4)\%$, in these experimental conditions of observation.

First-order radiative corrections with peaking approximation [1], when applied to the reaction

$$e^+e^- \to e^\pm e^\mp , \qquad (1)$$

predict that the final-state electron-positron pair could have a large acollinearity angle, but be in a plane containing the colliding beam axis.

We have studied reaction (1) using the Frascati colliding beam facility ADONE for total c.m. energies from 1.6 to 2.0 GeV, and for scattering angles from 45° to 135°. The experimental apparatus has been described elsewhere [2].

In the one-photon approximation, reaction (1) is described by the two Feynman diagrams and by their interference:

The first-order radiative corrections with peaking approximation essentially consist in adding to any one of the four electronic legs, a photon whose direction is assumed to be along the electron or positron line of flight. Thus, photons emitted in the final state do not destroy the collinearity of the pair; photons emitted in the initial state impart a non-zero velocity to the e^+e^- centre-of-mass system, which results in the angular distribution properties stated above.

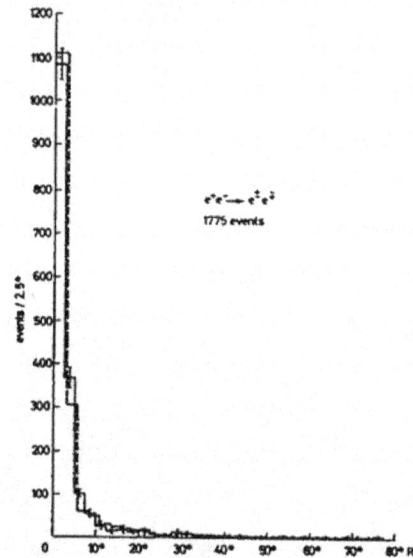

Fig. 1. Collinearity distribution integrated over all energies. The dotted histogram is the theoretical prediction based on the first-order radiative corrections with peaking approximation.

$e^+e^- \to e^\pm e^\mp$
1775 events

Fig. 6.2a: Front page of ref. 22 on the first observation of acoplanar radiative effects at ADONE.

It is interesting to read what he says: "Fortunately these problems can be overcome in the newly developed electron-positron colliding-beam accelerators where charged leptons can be copiously produced through the process[5]

$$e^+e^- \rightarrow \mu'^+ + \mu'^-$$

Within five years, through this process, we shall know if the electron-muon family has additional members with masses in the several-GeV ranges". ($\mu' \equiv HL$ of BCF group in Reference 5, i.e. Reference 19 of the present report).

In 1971 we published our results on the first observation of acoplanar radiative effects [22]. Figure 6.2a reproduces the front page of our paper, and Figure 6.2b the page where we established the effect to be (2.8 ± 0.4) % in our experimental conditions.

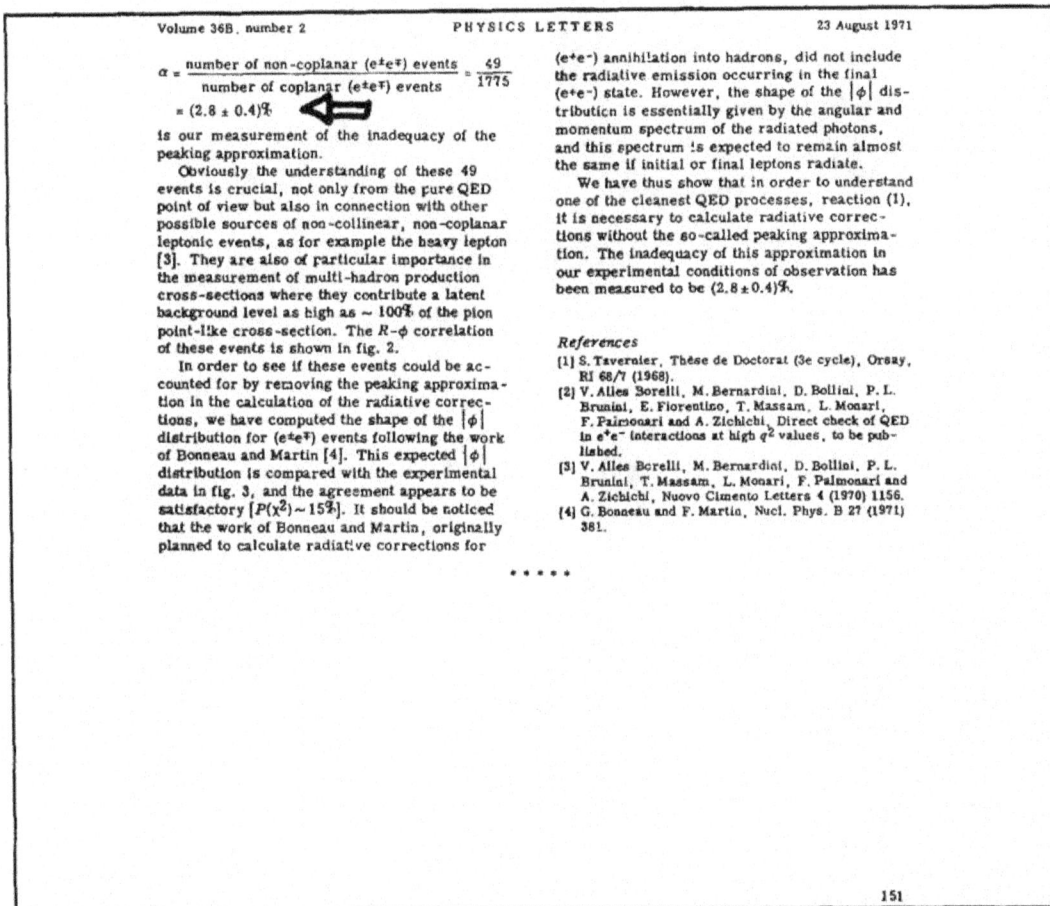

Volume 36B, number 2   PHYSICS LETTERS   23 August 1971

$$\alpha = \frac{\text{number of non-coplanar } (e^\pm e^\mp) \text{ events}}{\text{number of coplanar } (e^\pm e^\mp) \text{ events}} = \frac{49}{1775}$$

$$= (2.8 \pm 0.4)\%$$

is our measurement of the inadequacy of the peaking approximation.

Obviously the understanding of these 49 events is crucial, not only from the pure QED point of view but also in connection with other possible sources of non-collinear, non-coplanar leptonic events, as for example the heavy lepton [3]. They are also of particular importance in the measurement of multi-hadron production cross-sections where they contribute a latent background level as high as ~ 100% of the pion point-like cross-section. The $R-\phi$ correlation of these events is shown in fig. 2.

In order to see if these events could be accounted for by removing the peaking approximation in the calculation of the radiative corrections, we have computed the shape of the $|\phi|$ distribution for $(e^\pm e^\mp)$ events following the work of Bonneau and Martin [4]. This expected $|\phi|$ distribution is compared with the experimental data in fig. 3, and the agreement appears to be satisfactory $[P(\chi^2) \sim 15\%]$. It should be noticed that the work of Bonneau and Martin, originally planned to calculate radiative corrections for

(e^+e^-) annihilation into hadrons, did not include the radiative emission occurring in the final (e^+e^-) state. However, the shape of the $|\phi|$ distribution is essentially given by the angular and momentum spectrum of the radiated photons, and this spectrum is expected to remain almost the same if initial or final leptons radiate.

We have thus show that in order to understand one of the cleanest QED processes, reaction (1), it is necessary to calculate radiative corrections without the so-called peaking approximation. The inadequacy of this approximation in our experimental conditions of observation has been measured to be $(2.8 \pm 0.4)\%$.

References

[1] S. Tavernier, Thèse de Doctorat (3e cycle), Orsay, RI 68/7 (1968).

[2] V. Alles Borelli, M. Bernardini, D. Bollini, P. L. Brunini, E. Fiorentino, T. Massam, L. Monari, F. Palmonari and A. Zichichi, Direct check of QED in e^+e^- interactions at high q^2 values, to be published.

[3] V. Alles Borelli, M. Bernardini, D. Bollini, P. L. Brunini, T. Massam, L. Monari, F. Palmonari and A. Zichichi, Nuovo Cimento Letters 4 (1970) 1156.

[4] G. Bonneau and F. Martin, Nucl. Phys. B 27 (1971) 381.

· · · · ·

151

Fig. 6.2b: The page of ref. 22 where the relevant quantitative result on acoplanar radiative effects measured at ADONE is given.

We had also checked the validity of crossing symmetry in QED [21], as shown in Figure 6.3.

LETTERE AL NUOVO CIMENTO   VOL. 2, N. 7   14 Agosto 1971

Experimental Check of Crossing Symmetry in the Electromagnetic Interaction of Leptons.

V. ALLES BORELLI, M. BERNARDINI, D. BOLLINI, P. L. BRUNINI, E. FIORENTINO, T. MASSAM, L. MONARI, F. PALMONARI and A. ZICHICHI

CERN - Geneva
Istituto Nazionale di Fisica Nucleare - Sezione di Bologna
Istituto di Fisica dell'Università - Bologna
Laboratori Nazionali - Frascati

(ricevuto il 2 Luglio 1971)

A fundamental theorem of quantum field theory is crossing symmetry [1]. QED being the only working example of field theory, it is indeed crossing symmetric.

A straightforward check of QED crossing symmetry would be possible through a comparison between timelike and spacelike lepton-photon processes. In the one-photon approximation, this is shown in the diagrams below, where L stands for either electron or muon.

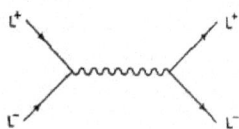

I) Timelike diagram.   II) Spacelike diagram.

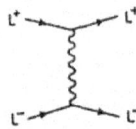

If we call $F^{LL\gamma}(q^2)$ the vertex function which describes the electromagnetic interaction between the lepton and the photon, crossing symmetry says that this vertex function is the same analytic function for timelike and spacelike processes, the only change being the value of the variable q^2.

The experimental check we propose for this QED crossing symmetry is based on comparison of the following two leptonic processes:

(1) $$e^+e^- \to e^\pm e^\mp ,$$

(2) $$e^+e^- \to \mu^\pm \mu^\mp ,$$

which have been studied at Frascati using the colliding-beam facility Adone.

[1] M. GELL-MANN and M. L. GOLDBERGER: *Phys. Rev.*, **96**, 1433 (1954).

Fig. 6.3:   Front page of ref. 21 on the study of QED crossing symmetry at ADONE.

IL NUOVO CIMENTO   VOL. 7 A, N. 2   21 Gennaio 1972

A Check of Quantum Electrodynamics and of Electron-Muon Equivalence.

V. Alles-Borelli, M. Bernardini, D. Bollini, P. L. Brunini,
E. Fiorentino, T. Massam, L. Monari, F. Palmonari and A. Zichichi

CERN - Geneva
Istituto Nazionale di Fisica Nucleare - Sezione di Bologna
Istituto di Fisica dell'Università - Bologna
Laboratori Nazionali - Frascati

(ricevuto il 28 Maggio 1971; manoscritto revisionato ricevuto l'8 Luglio 1971)

Summary. — A study of the timelike reaction $e^+e^- \to \mu^\pm\mu^\mp$ and of its comparison with the spacelike dominated reaction $e^+e^- \to e^\pm e^\mp$ allows us to establish the validity of QED in terms of production angular distributions, absolute rates, energy dependence and angular correlations between the pair of final-state leptons, in two very different ranges of invariant four-momentum transfer. No sign of QED break is detected in the electromagnetic interaction of leptons and the muon behaves like a heavy electron, within the accuracy of the present investigation.

1. – Introduction.

We report here a study of the timelike process

$$(1) \qquad\qquad e^+e^- \to \mu^\pm\mu^\mp$$

and its comparison with the spacelike dominated process

$$(2) \qquad\qquad e^+e^- \to e^\pm e^\mp .$$

330

Fig. 6.4: Front page of ref. 23 on the study of the electron-muon QED equivalence at ADONE.

Later on we published all our results on QED checks for electrons and muons as shown: in Figure 6.4 for the electron-muon QED equivalence [23]; in Figure 6.5 for a measurement, with ten times more statistics, of the acoplanar radiative effects for (e^+e^-) final states [25]; in Figures 6.6a and 6.6b for a high precision QED test of the energy dependence of $e^+e^- \to e^+e^-$ [24, 26]; in Figure 6.7 for the energy dependence of $e^+e^- \to \mu^+\mu^-$ [30]. The final result on the validity of the $(e \neq \mu)$ leptonic selection rule at the $7 \cdot 10^{-5}$ level is shown in Figure 6.8 [45], and on the study of the acoplanar $e^+e^- \to \pi^\pm\pi^\mp + X$ reaction [20] in Figure 6.9.

This hadronic channel was particularly relevant as a background for our sought after $(e^\pm\mu^\mp)$ acoplanar events.

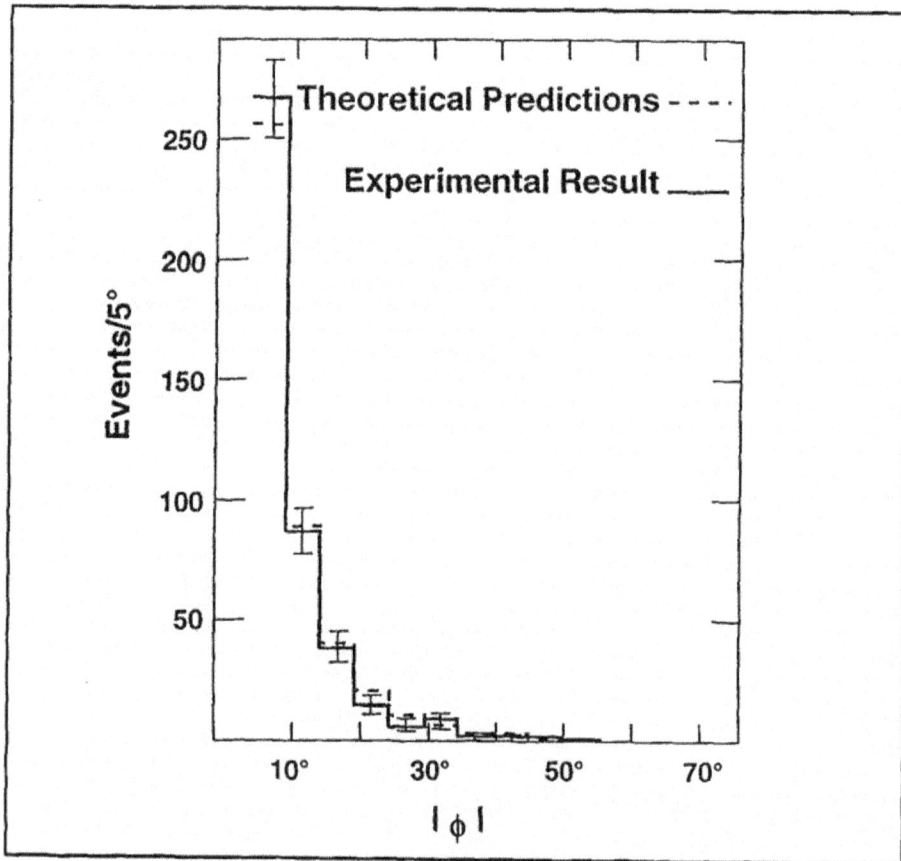

Fig. 6.5: Measurement of the acoplanar radiative effects for (e^+e^-) final states at ADONE [25], showing the acoplanarity angle (ϕ) distribution for 429 (e^+e^-) pairs with $|\phi| > 5°$.

APPENDIX A

CERN
SERVICE D'INFORMATION
SCIENTIFIQUE

Volume 45B, number 5   PHYSICS LETTERS   20 August 1973

ACCURATE MEASUREMENT OF THE ENERGY DEPENDENCE OF THE PROCESS
$e^+ + e^- \to e^\pm + e^\mp$, IN THE s-RANGE 1.44–9.0 GeV²

M. BERNARDINI, D. BOLLINI, P.L. BRUNINI, E. FIORENTINO,
T. MASSAM, L. MONARI, F. PALMONARI, F. RIMONDI and A. ZICHICHI

CERN, Geneva, Switzerland
Istituto Nazionale di Fisica Nucleare, Bologna, Italy
Istituto di Fisica dell 'Università, Bologna, Italy
and Laboratori Nazionali del CNEN, Frascati-Roma, Italy

Received 14 June 1973

The analysis of 12 827 $e^+ + e^- \to e^\pm + e^\mp$ events observed in the s-range 1.44–9.0 GeV² allows measurement of the energy dependence of the cross-section for the most typical QED process, with ±2% accuracy. Within this limit the data follow QED, with first-order radiative corrections included.

Using the Bologna-CERN set-up, the reaction

$$e^+ + e^- \to e^\pm + e^\mp \qquad (1)$$

has been studied at the ADONE colliding beam machine in Frascati. The purpose of this work was to check the validity of QED in the s-range 1.44–9.0 GeV².

The apparatus consisted of : i) thin-plate spark chambers of kinematic reconstruction; ii) heavy-plate spark chambers for particle identification (electrons make showers; muons show only Coulomb scattering; pions and kaons produce the typical hadronic patterns); iii) a system of plastic scintillation counters for accurate time-of-flight (±0.5 ns) and for other fast-trigger purposes. Details of the set-up have already been published [1, 2] and will not be repeated here.

The QED check we report here is based on a comparison between "large-angle" and "small-angle" data from reaction (1). The small-angle data are used for luminosity measurements and cover a very low t-range, with an average t-value over the small-angle telescope of -2×10^{-3} GeV² (spacelike). The theoretical implications of this study have already been discussed in a previous work [2]. Table 1 shows the luminosities, the total number of large-angle ($e^\pm e^\mp$) events, and the t-ranges, corresponding to the various energies investigated with the present work. The results obtained are based on the analysis of 12 827 events.

These events have been reconstructed in space and fully analysed in order to exclude background contamination from beam-gas interactions and cosmic

Table 1

E_{beam} (MeV)	L_I (nb⁻¹)	$N_{(e^\pm e^\mp)}$ $R \leqslant 10^\circ$ $|\Phi| < 5^\circ$	t-range (GeV²)
600	4.98	570	−(0.241 − 1.2)
650	8.02	665	−(0.283 − 1.41)
700	10.95	765	−(0.328 − 1.63)
750	27.15	1565	−(0.377 − 1.87)
800	10.23	460	−(0.429 − 2.13)
850	13.05	516	−(0.484 − 2.41)
950	60.95	1900	−(0.605 − 3.00)
970	23.61	668	−(0.623 − 3.13)
1050	186.09	4240	−(0.739 − 3.67)
1200	44.90	684	−(0.965 − 4.79)
1500	81.64	794	−(1.507 − 7.49)
Totals	431.57	12827	

rays. The background level of these sources has been shown to be negligible (see refs. [2] and [3] for a discussion of backgrounds). What is not negligible is the contribution due to first-order radiative corrections. These have been calculated following the work of Tavernier [4] and of Calva-Tellez [5]; so by comparing our results with the theoretical predictions we can establish if QED is valid to the third-order in the electromagnetic coupling.

† R is the deviation from collinearity; $R = 0$ means collinear event. The acoplanarity angle ϕ is the angle between the two planes, which contain the two particles of the final state in reaction (1) and the beam axis.

Fig. 6.6a: Front page of ref. 26 on the precision measurement of the energy dependence of $\sigma(e^+e^- \to e^+e^-)$ at ADONE.

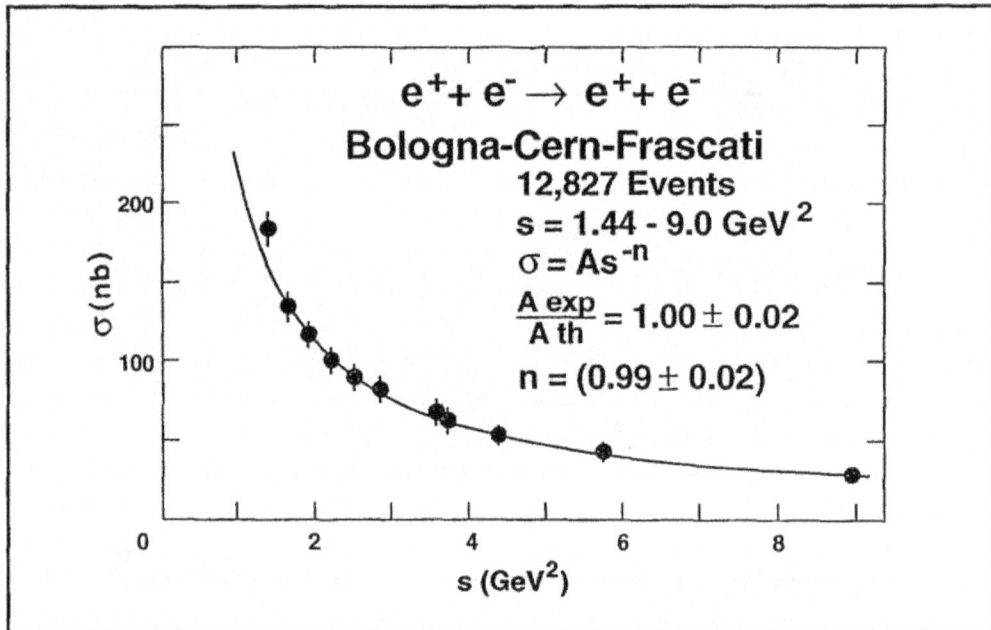

Fig. 6.6b: The cross section $\sigma\,(e^+e^- \to e^+e^-)$ vs. s, as measured at ADONE [26].

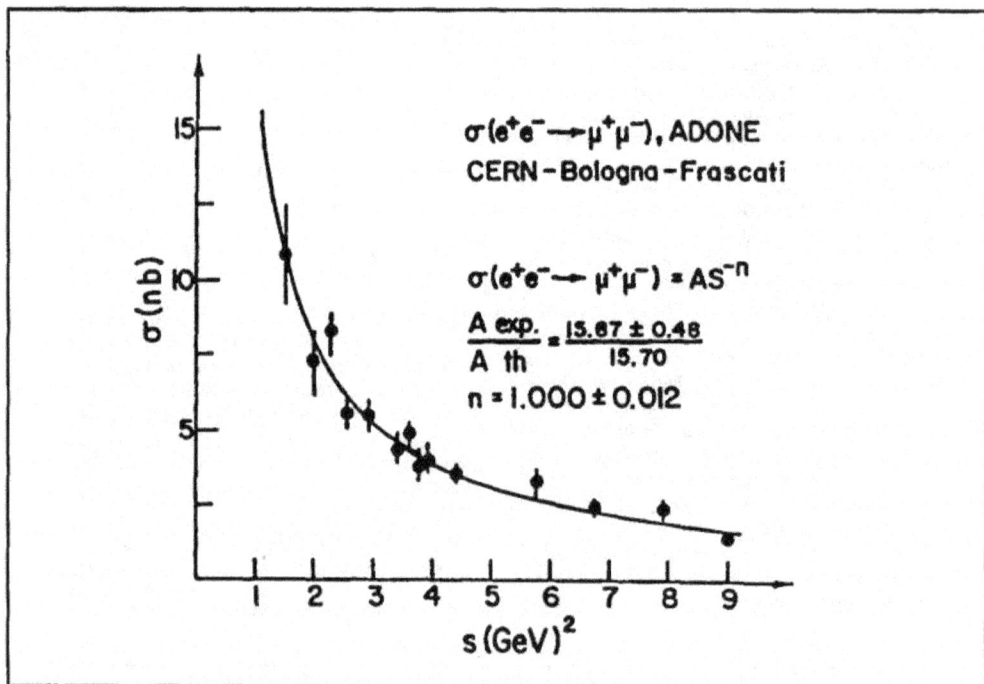

Fig. 6.7:   The cross section $\sigma\,(e^+e^- \to \mu^+\mu^-)$ vs. s, as measured at ADONE [30].

A. ZICHICHI

Now the point is experimental. So far no states with $C = +1$ have been observed, and the experimental investigation has just started with the study of the purely electromagnetic processes

$$e^+e^- \to e^+e^-e^+e^-,$$

$$e^+e^- \to e^+e^-\mu^+\mu^-,$$

observed at Novosibirsk [34] and Frascati [35].

There is a straightforward way [36] to detect $C = +1$ states; this is via the determination of the turning point in the production angular distribution. High-energy machines will certainly have the privilege of taking up this field of great interest for hadrodynamics.

9. – Validity of the leptonic selection rules.

In the field of weak interactions we know that the transition $\mu \leftrightarrow e$ is forbidden by the known leptonic selection rules. The available experimental information does not allow a distinction to be made between the two alternative classes of selection rules [37] which distinguish the « electron world » from the « muon world »; namely: a) two additive selection rules, b) an additive and a multiplicative selection rule. Both sets of rules would be violated by the existence of the process

$$e^+e^- \to \mu^\pm e^\mp.$$

For very low q^2 values ($q^2 \simeq 0.01$ (GeV)^2) it is known that the transition $\mu \leftrightarrow e$ is strongly depressed. Examples are the unobserved processes

$$\mu^\pm \to e^\pm + \gamma,$$

$$\mu^- + \text{nucleus} \to (\text{nucleus})' + e^-.$$

However, as nobody knows the reason for the existence of these two leptonic quantum numbers, it is of interest to study their validity *vs.* q^2. For example, in the Pati-Salam [38] theory with the leptonic number taken as the 4th colour, there is no place for this selection rule, a way out being its violation at some value of q^2. We have looked for collinear ($\mu^\pm e^\mp$) pairs in our sample of events, where $\sim 14\,000$ (e^+e^-) and 1000 ($\mu^+\mu^-$) were observed. The ranges of s and of t are given in Table II. From the total number of observed lepton pairs, $N(\text{lepton} + \text{antilepton}) = 15\,000$, we obtain the following limit for the branching ratio:

$$\frac{e^+e^- \to \mu^\pm e^\mp}{e^+e^- \to \text{lepton} + \text{antilepton}} < 7 \cdot 10^{-6}.$$

Fig. 6.8: Page of ref. 45 where the final, high-statistics limit obtained at ADONE on the branching ratio ($e^+e^- \to \mu^\pm e^\mp / e^+e^- \to l\,\bar{l}$), with $l = e, \mu$, is reported.

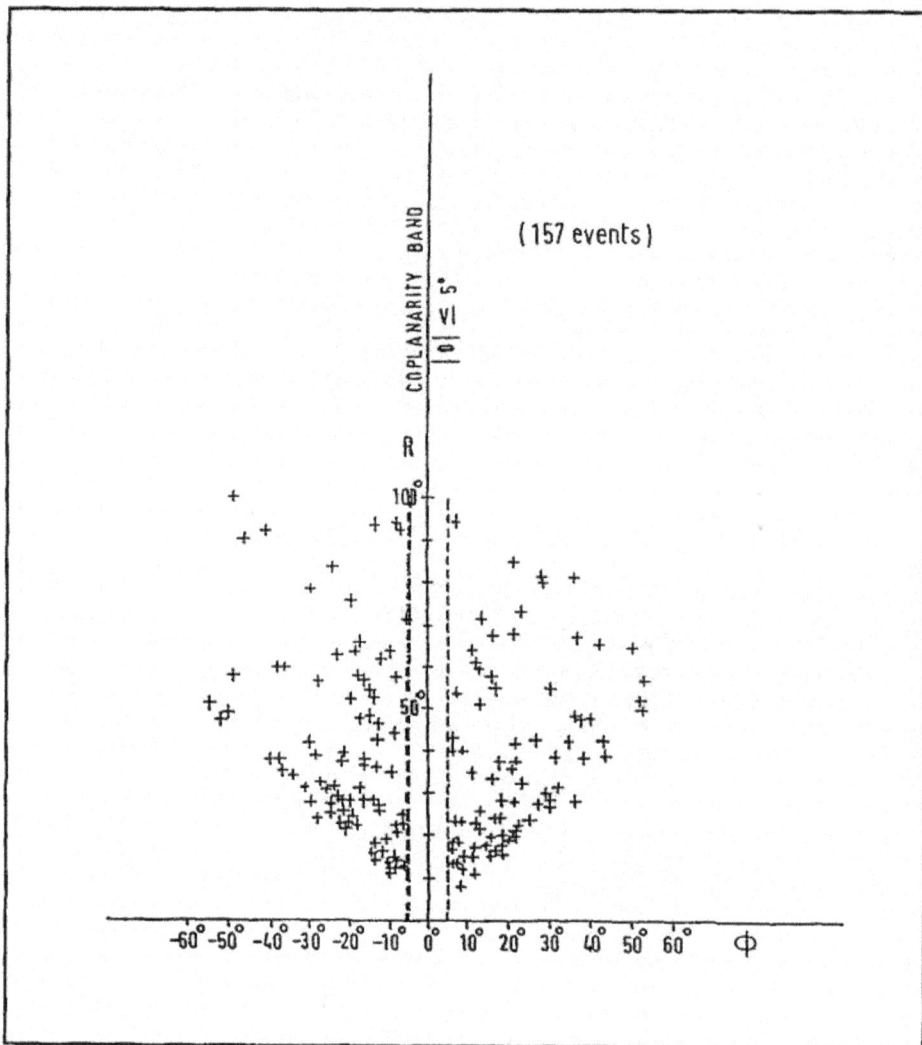

Fig. 6.9: R (acollinearity angle) vs. ϕ (acoplanarity angle) scatter diagram for $(\pi^{\pm}\pi^{\mp})$ outside the coplanar band (i.e. $|\phi| > 5°$), as observed at ADONE [20]. The non-collinear, non-coplanar events correspond to $e^+e^- \to \pi^{\pm}\pi^{\mp} +$ anything.

By the time we concluded our experiment at Frascati, the new class of Heavy Leptons, E and M, had been proposed by theorists, as we have reported in our final Frascati paper [27], whose first page is reproduced in Figure 6.10. This is in fact our 1973 final result on the limits of the mass of the Heavy Lepton via the study of acoplanar $(e^{\pm}\mu^{\mp})$ pairs. In Figure 6.11 the details on the ADONE energies and the corresponding integrated luminosities are given.

IL NUOVO CIMENTO   Vol. 17 A, N. 2   21 Settembre 1973

Limits of the Mass of Heavy Leptons.

M. Bernardini, D. Bollini, P. L. Brunini, E. Fiorentino,
T. Massam, L. Monari, F. Palmonari, F. Rimondi and A. Zichichi

CERN - Geneva
Istituto Nazionale di Fisica Nucleare - Sezione di Bologna
Istituto di Fisica dell'Università - Bologna
Laboratori Nazionale del CNEN - Frascati (Roma)

(ricevuto il 9 Luglio 1973)

Summary. — A further search for heavy leptons at the ADONE e^+e^- storage ring has revealed no events. This establishes, with 95% confidence, that, if a heavy lepton exists and is universally coupled only to ordinary leptons, its mass must be heavier than 1.4 GeV. If it is also coupled to the hadrons, its mass must be greater than 1 GeV, again with 95% confidence.

Fig. 6.10: Front page of ref. 27 where the final results on the Heavy Lepton search by the BCF group at ADONE are reported.

Limits on the Mass of Heavy Leptons

Table I

Beam Energy (MeV)	Integrated Luminosity (x 10^{32} cm^{-2})
600	50
650	80
700	74
750	175
800	102
850	130
950	630
970	235
1050	1861
1200	449
1500	800

If the heavy lepton is coupled only to ordinary leptons
(with the universal weak coupling constant),
the 95% confidence level for the mass is:

$$m_{HL} \geq 1.45 \text{ GeV}$$

If the heavy lepton is universally coupled to both
ordinary leptons and hadrons,
the 95% confidence level for the mass is:

$$m_{HL} \geq 1.0 \text{ GeV}$$

Fig. 6.11: Details on the ADONE energy and integrated luminosity used to derive the final results reported in ref. 27 on the Heavy Lepton mass limits.

In Figure 6.12 the Frascati results are presented in a graphical synthesis: for a Heavy Lepton having its own neutrino and being universally coupled with ordinary leptons and hadrons the mass limit is (at 95% CL) 1.0 GeV/c^2.

After the final results by the BCF group were published in 1973 [27], the proposal for further searches at energies higher than ADONE was promoted in a series of conference reports: at the 1972 EPS Wiesbaden Conference, at the 1973 Pavia Symposium, at the 1973 Frascati Meeting and at the 1973 Bielefeld International Discussion Meeting.

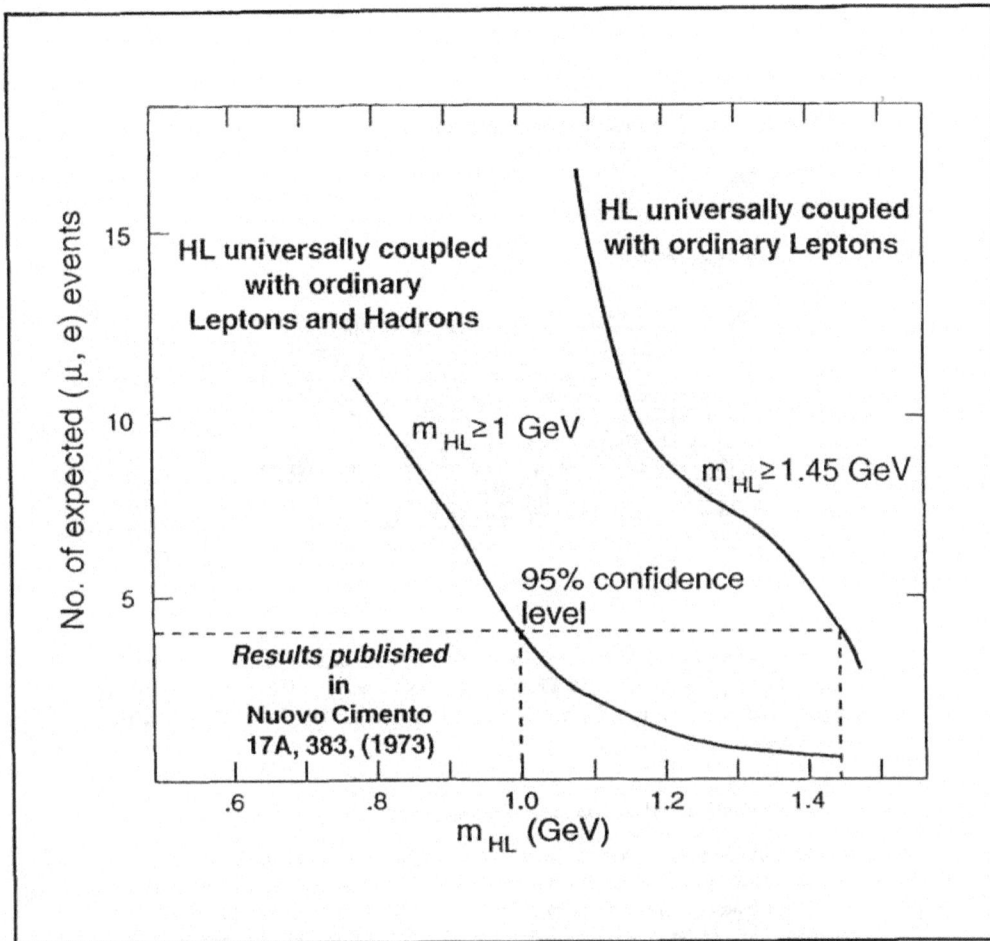

Fig. 6.12: The expected number of $(e^\pm\mu^\mp)$ pairs vs. m_{HL}, i.e. the Heavy Lepton mass, for two types of universal weak couplings of the Heavy Lepton [27].

I decided to publish a synthesis of all these reports, with the title "Why (e^+e^-) physics is fascinating" [45]. In this review paper, whose first page is shown in Figure 6.13, I discuss the newly "theoretically wanted" Heavy Leptons. In Figure 6.14 the relevant page with what I say of these "theoretically wanted Heavy Leptons" (already described in Section 3 of the present report) is reproduced. The pages of my (e^+e^-) synthesis, where I discuss the importance of studying decay correlations, decay spectra and decay rates in order to disentangle the Heavy Lepton from the "theoretically wanted" ones, are shown in Figures 6.15a and 6.15b.

RIVISTA DEL NUOVO CIMENTO   VOL. 4, N. 4   Ottobre-Dicembre 1974

Why (e⁺e⁻) Physics is Fascinating (*).

A. ZICHICHI (**)

CERN - Geneva

(ricevuto il 12 Aprile 1974)

Summary

The results obtained by the Bologna-CERN-Frascati Collaboration during about three years of work at Frascati are reviewed and taken as a basis to show the impact of (e⁺e⁻) physics in understanding the laws of subnuclear phenomena.

1. – Introduction: (e⁺e⁻) machines in the world.

At this Conference you have heard how the basic laws of hadrodynamics can be investigated when the initial state of a reaction consists of hadrons.

The purpose of this talk is to show what we can learn when the initial state consists only of leptons, and more precisely of a lepton-antilepton pair. The

(*) This paper is an updated version of two unpublished invited review papers presented at the EPS Conference, Wiesbaden, 3-6 October 1972, and at the IV International Symposium on Multiparticle Hadrodynamics, Pavia, 31 August-4 September 1973. The data on σ(e⁺e⁻ → hadrons) have been presented at: i) the XVI International Conference on High-Energy Physics, Batavia, Ill., 1972; ii) the Informal Meeting on Recent Developments in High-Energy Physics, Frascati, 26-31 March 1973; iii) the International Discussion Meeting on (e⁺e⁻) Annihilation, Bielefeld, 19-21 September 1973.
(**) On leave of absence from the University of Bologna.

Fig. 6.13: Front page of ref. 45, i.e. the final review paper on all the results obtained by the BCF group on (e⁺e⁻) collider physics at ADONE.

508   A. ZICHICHI

From the data on the reaction

$$e^+ + e^- \rightarrow \mu^+ + \mu^-$$

we conclude that a timelike vertex behaves like a spacelike vertex within $\pm 1\%$, as implied by crossing symmetry, a basic theorem of local relativistic quantum field theory [2].

We can close this Section by stating that so far we have every reason to believe that local relativistic quantum field theory is indeed an excellent tool for describing processes where leptons are involved. This is why this tool can be used for the more complex phenomena involving hadrons.

3. – Is it possible to renormalize weak interactions? Other heavy leptons?

A word of caution is needed here. For a long time physicists have been puzzled by the existence of only two types of leptons: the electronlike and the muonlike. When these are compared with the enormous variety of hadrons, the call for leptons outside the electron and the muon class becomes very natural. Such leptons have nothing to do with the leptons required by the gauge theories.

Fig. 6.14: The page of ref. 45 where the relevant point about "theoretically wanted" Heavy Leptons is outlined.

statistics and taking into account the fact that the calculated background and the observed number of events were both equal to 2.

If the heavy lepton is coupled only to ordinary leptons (with the universal weak-coupling constant), the 95% confidence level for the mass is

$$m_{HL} > 1.45 \text{ GeV} .$$

If the heavy lepton is universally coupled to both ordinary leptons and hadrons, then the 95% confidence level for the mass is

$$m_{HL} > 1.0 \text{ GeV} .$$

Notice that the above mass limits apply to any type of heavy lepton (E, M, L, t) quoted above. The present investigation thus establishes that neither heavy leptons required by the gauge theories of weak interactions nor heavy leptons of the old standard type exist with masses below 1.0 GeV.

Nevertheless, colliding (e⁺e⁻) beams will remain a very clean tool with which to search for heavy leptons. It is perhaps interesting to recall that if (e⁺μ∓) pairs were observed, their origin in terms of E, M, L could finally be established because there are three sources of information, which differ for the various leptons, *viz.* decay correlations, decay spectra, decay rates. For example, let us consider the decay of L⁺ and E⁺ (the double arrow indicates spin, a single arrow indicates momentum).

➡ *Decay correlations.*

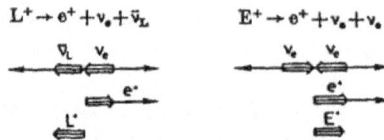

$$L^+ \to e^+ + \nu_e + \bar{\nu}_L \qquad\qquad E^+ \to e^+ + \nu_e + \nu_e .$$

Owing to the different neutrinos emitted in the decay of L⁺ and E⁺, the positive electron will be emitted in the opposite direction with respect to the L⁺ spin, but along the E⁺ spin, respectively. This is an interesting source for decay correlations.

➡ *Decay spectra.* The highest-energy configuration for e⁺ is allowed in the case of L⁺ decay and forbidden for E⁺ decay:

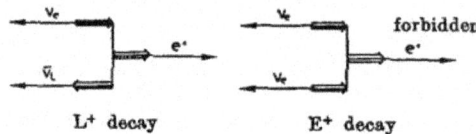

L⁺ decay   E⁺ decay

Fig. 6.15a: Showing the main items presented in ref. 45, i.e. the importance of studying decay correlations and decay spectra to identify the new Heavy Leptons.

The Michel spectrum will obviously be different because the highest-energy configuration must vanish for E^+ decays:

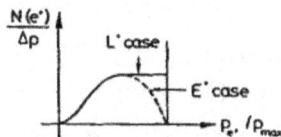

▷ *Decay rates.* The decay rates for the purely leptonic channels compare as follows:

$$\text{rate } (L^+ \to e^+ \nu_e \bar{\nu}_L) = \text{rate } (L^+ \to \mu^+ \nu_\mu \bar{\nu}_L) = \text{rate } (E^+ \to \mu^+ \nu_\mu \nu_e) = \text{rate } (M^+ \to e^+ \nu_e \nu_\mu) \,.$$

But

$$\text{rate } (E^+ \to e^+ \nu_e \nu_e) = 2 \ \text{rate } (E^+ \to \mu^+ \nu_\mu \nu_e) \,,$$

$$\text{rate } (M^+ \to \mu^+ \nu_\mu \nu_\mu) = 2 \ \text{rate } (M^+ \to e^+ \nu_e \nu_\mu) \,.$$

To see the origin of this factor of 2, let us take as example the E^+ decay; there are two diagrams contributing

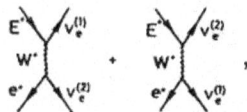

while for the μ^+-channel there is only one diagram

because the two neutrinos in the final states are not equal.

When compared to the μ^+-channel, the rate of the e^+-channel is therefore 2^2 (because of the two amplitudes obtained by interchanging the two identical neutrinos, labelled $\nu_e^{(1)}$ and $\nu_e^{(2)}$ in the diagram above), times one-half, because of the two identical particles in the final state, *i.e.*

$$\text{rate } (E^+ \to e^+ \nu_e \nu_e) = 2^2 \cdot \tfrac{1}{2} \cdot \text{rate } (E^+ \to \mu^+ \nu_\mu \nu_e) \,,$$

which is the result quoted above.

Fig. 6.15b: Same as Fig. 6.15a, concerning the *HL* decay rates for the purely leptonic channels [45].

7 — Summary and Conclusions.

Let me summarize this report and conclude.

I have described the foundations of Heavy Lepton searches where the key idea was the existence of a new Heavy Lepton carrying its own leptonic number. The time sequence was as follows.

The period 1960-1968 covered a series of published papers describing the experimental work at CERN dedicated to the search for time-like photons able to produce Heavy Lepton pairs carrying a new leptonic number and thus producing the $(e^{\pm}\mu^{\mp})$ signature. This was possible thanks to a powerful set-up able to simultaneously detect $(e^{\pm}e^{\mp})$, $(\mu^{\pm}\mu^{\mp})$ and $(e^{\pm}\mu^{\mp})$ pairs in hadronic processes. The technologically innovative key-point of this work was the invention of the "pre-shower" technique published in 1963. Notice that a large solid-angle detector able to observe $(e^{\pm}\mu^{\mp})$ pairs produced in hadronic interactions was fully operative at CERN already in 1964.

From 1963 to 1970 the technology was transferred from CERN to Frascati. In 1967 the BCF group presented its proposal to INFN. All the calibration work (Time-Of-Flight, heavy-plate spark chambers, π/e and π/μ) was performed at CERN with known beams of (e, μ, π) at the correct energies simulating ADONE final states. Without this extensive series of experimental studies at CERN it would have been impossible to search for a new Heavy Lepton in Frascati. The experimental set-up mounted around the ADONE interaction region was working as expected and the quality of the $(e^{+}e^{-})$ and $(\mu^{+}\mu^{-})$ final states reported in Figures 5.7 and 5.8 is an example. The selection power of the Frascati set-up was so good that we could determine, as by-product of our experiment, the EM form factor of the pseudoscalar mesons $(\pi$ and $K)$ in the time-like range allowed by the ADONE energies. Our set-up was in fact conceived to distinguish clearly between all known particles: electrons, muons, pions and kaons.

The time interval 1967-1970 is characterized by the INFN proposal and the first results.

The period 1971-1975 represents the crucial transition of the *HL* and the acoplanar $(e\mu)$ method, from Frascati to SLAC.

Let me now conclude by listing the following basic steps:

- the idea of a new lepton heavier than the known ones (e, μ) and carrying its own leptonic number;
- the choice of the best production process: $e^{+}e^{-} \rightarrow HL^{+}HL^{-}$;
- the invention of the technology to detect its existence: the acoplanar $(e^{\pm}\mu^{\mp})$ method;
- the implementation of the large solid angle detector needed to establish the first upper limit on the *HL* mass;

- the proof that the acoplanar $(e^{\pm}\mu^{\mp})$ method worked as expected in the design proposal;
- the promotion for the *HL* search at energies higher than ADONE.

The above points are the original contributions during more than a decade (1960-1975) of the PAPLEP (CERN)–BCF group to the discovery of the Heavy Lepton. Looking back, it gives me great pleasure to know that each step in this long search did play an important role in reaching the final goal.

8 — References.

[1]   *Evidence for anomalous lepton production in e^{+}-e^{-} annihilation*
        M.L. Perl et al.
        Physical Review Letters, 35, 1489 (1975).

[2]   *A proposal to search for leptonic quarks and heavy leptons produced by ADONE*
        M. Bernardini, D. Bollini, E. Fiorentino, F. Mainardi, T. Massam, L. Monari,
        F. Palmonari and A. Zichichi.
        INFN/AE-67/3, 20 March 1967.

[3]   *Measurement of the anomalous magnetic moment of the muon*
        G. Charpak, F.J.M. Farley, R.L. Garwin, T. Muller, J.C. Sens, V.L. Telegdi and
        A. Zichichi.
        Physical Review Letters, 6, 128 (1961).

[4]   *A telescope to identify electrons in the presence of pion background*
        T. Massam, Th. Muller and A. Zichichi.
        CERN 63-25, 27 June 1963.

[5]   *A new electron detector with high rejection power against pions*
        T. Massam, Th. Muller, M. Schneegans and A. Zichichi.
        Nuovo Cimento, 39, 464 (1965).

[6]   *Un grand détecteur E.M. à haute réjection des pions*
        D. Bollini, A. Buhler-Broglin, P. Dalpiaz, T. Massam, F. Navach, F.L. Navarria,
        M.A. Schneegans and A. Zichichi.
        Revue·de Physique Appliquée, 4, 108 (1969).

[7]   *A large electromagnetic shower detector with high rejection power against pions*
        M. Basile, J. Berbiers, D. Bollini, A. Buhler-Broglin, P. Dalpiaz, P.L. Frabetti,
        T. Massam, F. Navach, F.L. Navarria, M.A. Schneegans and A. Zichichi.
        Nuclear Instruments and Methods, 101, 433 (1972).

[8]   *Range measurements for muons in the GeV region*
        A. Buhler, T. Massam, Th. Muller and A. Zichichi.
        CERN 64-31, 24 June 1964.

[9]   *Range measurements for muons in the GeV region*
        A. Buhler, T. Massam, Th. Muller and A. Zichichi.
        Nuovo Cimento, 35, 759 (1965).

[10]   *A new large-acceptance and high-efficiency neutron detector for missing-mass studies*
        D. Bollini, A. Buhler-Broglin, P. Dalpiaz, T. Massam, F. Navach, F.L. Navarria,
        M.A. Schneegans, F. Zetti and A. Zichichi.
        Nuovo Cimento, 61A, 125 (1969).

[11]   *Search for the time-like structure of the proton*
        M. Conversi, T. Massam, Th. Muller and A. Zichichi.
        Physics Letters, 5, 195 (1963).

[12]   *An experiment on the time-like electromagnetic structure of the proton*
        M. Conversi, T. Massam, Th. Muller and A. Zichichi.
        Proceedings of the International Conference on "Elementary Particles", Siena, Italy, 30
        September-5 October 1963 (Soc. Ital. di Fisica, Bologna, 1964), Vol. 1, 488.

[13]   *Proton antiproton annihilation into muon pair*
        M. Conversi, T. Massam, Th. Muller, M. Schneegans and A. Zichichi.
        Proceedings of the International Conference on "High-Energy Physics", Dubna,
        USSR, 5-15 August 1964 (Atomizdat, Moscow, 1966), Vol. I, 857.

[14]   *The leptonic annihilation modes of the proton-antiproton system at 6.8 (GeV/c)^2*
        timelike four-momentum transfer
        M. Conversi, T. Massam, Th. Muller and A. Zichichi.
        Nuovo Cimento, 40, 690 (1965).

[15]   *Observation of the rare decay mode of the ϕ-meson: $\phi \to e^+e^-$*
        D. Bollini, A. Buhler-Broglin, P. Dalpiaz, T. Massam, F. Navach, F.L. Navarria,
        M.A. Schneegans and A. Zichichi.
        Nuovo Cimento, 56A, 1173 (1968).

[16]   *The decay mode $\omega \to e^+e^-$ and a direct determination of the ω–ϕ mixing angle*
        D. Bollini, A. Buhler-Broglin, P. Dalpiaz, T. Massam, F. Navach, F.L. Navarria,
        M.A. Schneegans and A. Zichichi.
        Nuovo Cimento, 57A, 404 (1968).

[17]   *The basic SU(3) mixing: $\omega_8 \longleftrightarrow \omega_1$*
        A. Zichichi.
        "Evolution of Particle Physics" (Academic Press Inc., New York-London, 1970), 299.

[18]   *Validity of the leptonic selection rules for the ($\mu e \gamma$) vertex at high four-momentum transfers*
V. Alles-Borelli, M. Bernardini, D. Bollini, P.L. Brunini, T. Massam, L. Monari, F. Palmonari and A. Zichichi.
Lettere al Nuovo Cimento, 4, 1151 (1970).

[19]   *Limits on the electromagnetic production of heavy leptons*
V. Alles-Borelli, M. Bernardini, D. Bollini, P.L. Brunini, T. Massam, L. Monari, F. Palmonari and A. Zichichi.
Lettere al Nuovo Cimento, 4, 1156 (1970).

[20]   *Study of charged final states produced in e^+e^- interactions*
V. Alles-Borelli, M. Bernardini, D. Bollini, P.L. Brunini, E. Fiorentino, P.L. Frabetti, T. Massam, L. Monari, F. Palmonari and A. Zichichi.
Proceedings of the VIII Course of the "Ettore Majorana" International School of Subnuclear Physics, Erice, Italy, 1970: "Elementary Processes at High Energy" (Academic Press Inc., New York-London, 1971), 790.

[21]   *Experimental check of crossing symmetry in the electromagnetic interaction of leptons*
V. Alles-Borelli, M. Bernardini, D. Bollini, P.L. Brunini, E. Fiorentino, T. Massam, L. Monari, F. Palmonari and A. Zichichi.
Lettere al Nuovo Cimento, 2, 376 (1971).

[22]   *Experimental proof of the inadequacy of the peaking approximation in radiative corrections*
V. Alles-Borelli, M. Bernardini, D. Bollini, P.L. Brunini, E. Fiorentino, T. Massam, L. Monari, F. Palmonari and A. Zichichi.
Physics Letters, 36B, 149 (1971).

[23]   *A check of quantum electrodynamics and of electron-muon equivalence*
V. Alles-Borelli, M. Bernardini, D. Bollini, P.L. Brunini, E. Fiorentino, T. Massam, L. Monari, F. Palmonari and A. Zichichi.
Nuovo Cimento, 7A, 330 (1972).

[24]   *Direct check of QED in e^+e^- interactions at high q^2-values*
V. Alles-Borelli, M. Bernardini, D. Bollini, P.L. Brunini, E. Fiorentino, T. Massam, L. Monari, F. Palmonari and A. Zichichi.
Nuovo Cimento, 7A, 345 (1972).

[25]   *Acoplanar (e^+e^-) pairs and radiative corrections*
M. Bernardini, D. Bollini, P.L. Brunini, E. Fiorentino, T. Massam, L. Monari, F. Palmonari, F. Rimondi and A. Zichichi.
Physics Letters, 45B, 169 (1973).

[26] *Accurate measurement of the energy dependence of the process $e^+ + e^- \rightarrow e^\pm + e^\mp$, in the s-range 1.44-9.0 GeV^2*
M. Bernardini, D. Bollini, P.L. Brunini, E. Fiorentino, T. Massam, L. Monari, F. Palmonari, F. Rimondi and A. Zichichi.
Physics Letters, 45B, 510 (1973).

[27] *Limits on the mass of heavy leptons*
M. Bernardini, D. Bollini, P.L. Brunini, E. Fiorentino, T. Massam, L. Monari, F. Palmonari, F. Rimondi and A. Zichichi.
Nuovo Cimento, 17A, 383 (1973).

[28] *A detailed study of exclusive and inclusive (e^+e^-) induced processes in the energy range 1.2-3.0 GeV*
M. Bernardini, D. Bollini, P.L. Brunini, T. Massam, L. Monari, F. Palmonari, F. Rimondi and A. Zichichi.
Proceedings of the 17th International Conference on "High-Energy Physics", London, UK, 1-10 July 1974 (Rutherford Lab., Chilton, 1974), IV-6.

[29] *An experimental study of acoplanar ($\mu^\pm \mu^\mp$) pairs produced in (e^+e^-) annihilation*
D. Bollini, P. Giusti, T. Massam, L. Monari, F. Palmonari, G. Valenti and A. Zichichi.
Lettere al Nuovo Cimento, 13, 380 (1975).

[30] *Measurements of $\sigma(e^+e^- \rightarrow \mu^\pm \mu^\mp)$ in the energy range 1.2-3.0 GeV*
V. Alles-Borelli, M. Bernardini, D. Bollini, P. Giusti, T. Massam, L. Monari, F. Palmonari, G. Valenti and A. Zichichi.
Physics Letters, 59B, 201 (1975).

[31] *Proof of hadron production in e^+e^- interactions*
V. Alles-Borelli, M. Bernardini, D. Bollini, P.L. Brunini, E. Fiorentino, T. Massam, L. Monari, F. Palmonari, G. Valenti and A. Zichichi.
Proceedings of the International Conference on "Meson Resonances and Related Electromagnetic Phenomena", Bologna, Italy, 14-16 April 1971 (Editrice Compositori, Bologna, 1972), 489.

[32] *e^+e^- annihilation into two hadrons in the energy interval 1400-2400 MeV*
V. Alles-Borelli, M. Bernardini, D. Bollini, P.L. Brunini, E. Fiorentino, T. Massam, L. Monari, F. Palmonari, F. Rimondi and A. Zichichi.
Physics Letters, 40B, 433 (1972).

[33] *Proof of comparable K-pair and π-pair production from time-like photons of 1.5, 1.6, and 1.7 GeV, and determination of the K-meson electromagnetic form factor*
M. Bernardini, D. Bollini, P.L. Brunini, E. Fiorentino, T. Massam, L. Monari, F. Palmonari, F. Rimondi and A. Zichichi.
Physics Letters, 44B, 393 (1973).

[34] *The time-like electromagnetic form factors of the charged pseudoscalar mesons from 1.44 to 9.0 GeV^2*
M. Bernardini, D. Bollini, P.L. Brunini, E. Fiorentino, T. Massam, L. Monari, F. Palmonari, F. Rimondi and A. Zichichi.
Physics Letters, 46B, 261 (1973).

[35] *The energy dependence of $\sigma(e^+e^- \rightarrow hadrons)$ in the total centre-of-mass energy 1.2 to 3.0 GeV*
M. Bernardini, D. Bollini, P.L. Brunini, E. Fiorentino, T. Massam, L. Monari, F. Palmonari, F. Rimondi and A Zichichi.
Physics Letters, 51B, 200 (1974).

[36] *Cross-section measurements for the exclusive reaction $e^+e^- \rightarrow 4\pi^\pm$ in the energy range 1.2-3.0 GeV*
M. Bernardini, D. Bollini, P.L. Brunini, E. Fiorentino, T. Massam, L. Monari, F. Palmonari, F. Rimondi and A. Zichichi.
Physics Letters, 53B, 384 (1974).

[37] *A study of the hadronic angular distribution in (e^+e^-) processes from 1.2 to 3.0 GeV*
M. Bernardini, D. Bollini, P.L. Brunini, E. Fiorentino, T. Massam, L. Monari, F. Palmonari, F. Rimondi and A. Zichichi.
Nuovo Cimento, 26A, 163 (1975).

[38] *The pion electromagnetic form factor in the timelike range (1.44-9.0) (GeV)^2*
D. Bollini, P. Giusti, T. Massam, L. Monari, F. Palmonari, G. Valenti and A. Zichichi.
Lettere al Nuovo Cimento, 14, 418 (1975).

[39] *The present knowledge of $|F_\pi|$ and the value of a low-energy (e^+e^-) storage ring*
M. Basile, D. Bollini, G. Cara Romeo, L. Cifarelli, P. Giusti, T. Massam, L. Monari, F. Palmonari, M. Placidi, G. Valenti and A. Zichichi.
Nuovo Cimento, 34A, 1 (1976).

[40] *The heartbeat of the proton*
V.F. Weisskopf.
in *"Lepton Physics at CERN and Frascati"*, Ed. N. Cabibbo, to appear.

[41] *Heavy electrons and muons*
F.E. Low.
Physical Review Letters, 14, 328 (1965).

[42]   *Search for new particles produced by high-energy photons*
       A. Barna, J. Cox, F. Martin, M.L. Perl, T.H. Tan, W.T. Toner, T.F. Zipf and
       E.H. Bellamy.
       Physical Review, 173, 1391 (1968).

[43]   For a review of the theoretically required heavy leptons with their production and decay
       properties, see for instance:
       Spontaneously broken gauge theories of weak interactions and heavy leptons
       J.D. Bjorken and Ch. Llewellyn Smith.
       Physical Review, 7D, 887 (1973).

[44]   *How does the muon differ from the electron?*
       M.L. Perl.
       Physics Today, July 1971, 34.

[45]   *Why (e^+e^-) physics is fascinating*
       A. Zichichi.
       La Rivista del Nuovo Cimento, 4, 498 (1974), based on the contribution to many
       conferences, in particular: the EPS Conference of Wiesbaden, 3-6 October 1972; and
       also the IV International Symposium on Multiparticle Hadrodynamics, Pavia, 31
       August-4 September 1973; the XVI International Conference on High-Energy Physics,
       Batavia, Ill., 1972; the Informal Meeting on Recent Developments in High-Energy
       Physics, Frascati, 26-31 March 1973; the International Discussion Meeting on (e^+e^-)
       Annihilation, Bielefeld, 19-21 September 1973.

Antonino Zichichi

ANSWERS TO CLAIMS FOR PRIORITIES ON HL AND TO CRITICISMS OF THE BCF EXPERIMENTAL SET-UP

Academy of Sciences - Bologna, Italy
CERN - Geneva, Switzerland
INFN - Bologna, Italy
University of Bologna, Italy

ANSWERS TO CLAIMS FOR PRIORITIES ON HL AND TO CRITICISMS OF THE BCF EXPERIMENTAL SET-UP

Antonino Zichichi

Table of Contents

ANSWERS TO CLAIMS FOR PRIORITIES ON HL AND
TO CRITICISMS OF THE BCF EXPERIMENTAL SET-UP

Antonino Zichichi

Academy of Sciences - Bologna, Italy
CERN - Geneva, Switzerland
INFN - Bologna, Italy
University of Bologna, Italy

1 —   Introduction.

To come up with original ideas has always been the way to achieve progress in physics, and in 1975, when the announcement came from SLAC that the existence of acoplanar $(e\mu)$ pairs had finally been detected experimentally by M. Perl, I was extremely happy. This announcement was the crowning achievement of more than a decade of continuous hard work which I had invested, together with all my collaborators, first to think of a new heavy lepton (HL) and work out all consequences, then to devise the instrumentation necessary to confirm its existence experimentally.

I could not help but notice, however, that M. Perl was not giving my group and myself due credit. Nor did he clearly acknowledge that we were the first to:

- elaborate the concept of the new heavy lepton carrying its own leptonic number and coupled to its own neutrino (HL);
- run the first series of experiments for its detection from 1960 to 1964 at CERN;
- come up with the correct conclusion about the most effective production process $[(e^+e^-)$ annihilation] and the best signature [acoplanar $(e\mu)$ pairs] for its existence, as was proved by the original set-up in Frascati for the experiments conducted from 1967 to 1973.

This attitude affected me negatively, but I decided not to react, being engaged in new experiments and confident that all my peers knew, beyond any doubt, the real origin of the HL.

Then came the time when a certain group of physicists decided that M. Perl needed to be considered sole and undisputed "father" of the HL. It suddenly became necessary to reject any claims I may have had. This is when rumours were circulated that "Zichichi is only one among many physicists who contributed to HL" ... "Others have had the idea of HL before Zichichi" ... "The set-up conceived by Zichichi in Frascati would not have allowed him to detect proof of the HL existence, regardless of the amount of energy available". The campaign was timely and quite well orchestrated. It fulfilled its purpose, and M. Perl was able to benefit from the a decade of hard work I and my collaborators had invested in the HL.

Needless to say, I was very upset at this demonstration of a rather extreme lobbying effort. However, the worst was still to come. Having reached their immediate objective, the rumours that were spread continued to linger and have lately reached the point where the past claims of my group on the HL started to be put in serious doubt. This is where the line had to be drawn. Integrity is by necessity a cornerstone of the world of science, and no-one in our community should be expected to tolerate such a distortion of the scientific truths.

The ten years of work for the Heavy Lepton are documented in my report to the Conference (Appendix A) but the answers to claims and criticisms are not. This Appendix is a synthesis of the answers presented in my review paper [1] which was prompted by the concern of some of my peers who convinced me of the necessity to place all events in their right perspective.

2 — Claims for priorities on HL.

In the literature of the years 1960-67, all sorts of proposals for all sorts of heavy leptons can be found, but none for HL. The published claims for precedence — or at least equal precedence — are all in M. Perl's report, presented at the International Conference on "The History of Original Ideas and Basic Discoveries in Particle Physics" [2]. Apart from claiming that he himself was thinking of a heavy lepton with the HL characteristics, Perl mentions Lipmanov [3], Cabibbo and Gatto [4] — referred to as "seminal" — , Zeldovich [5] and finally, Rothe and Wolsky [6]. It is worth noting that none of these were quoted or referred to by M. Perl in his 1971 Physics Today paper [7] when preparing for his SLAC experiment. Each of the above claims is discussed hereafter.

2.1 — M. Perl's 1963 claim.

Apart from it not being supported by any testimony or publication, it is difficult to believe that if Perl had indeed thought of a third lepton with the HL characteristics back in 1963, the experimental set-up at SLAC would have turned out, in the early seventies, to be so deficient in terms of $(e\mu)$ identification. Such an unsupported claim could never compare with the serious evidence represented by the testimonies of Victor F. Weisskopf [8], Nicola Cabibbo [9] and André Petermann [10]: a correct and complete concept of HL was conceived by A. Zichichi at CERN in 1960-61.

2.2 — Lipmanov.

The heavy lepton Lipmanov had proposed carried the same leptonic number as the known leptons e and μ and could not therefore be of the HL type.

2.3 — Cabibbo and Gatto.

The HL type of heavy lepton is not considered in their paper, and they themselves have never made any such a claim.

2.4 — Zeldovich.

Zeldovich discussed at length the importance of the conservation of two independent leptonic numbers, the "electron leptonic charge" and the "muon leptonic charge", neither of which have anything to do with the heavy lepton proposed and searched for at CERN and Frascati. He then devotes a few lines to higher mass leptons, where he fails to suggest new leptonic numbers or to refer to the properties of these particles and how to search for them. He even says that "this type of particle would be exceptionally difficult to observe". Nowhere in his work does Zeldovich mention the possibility of a third (or other) conserved lepton number, nor any particularly interesting final state to be observed. Such a "theoretical" view is not surprising, since the type of lepton proposed and searched for at CERN and Frascati fell outside the theoretical framework of the time.

2.5 — Rothe and Wolsky.

Their paper was mentioned by Perl as "indicative of the thinking on heavy leptons in the second half of the sixties". In the final states suggested by the authors, based on a theoretically incorrect scenario, "two oppositely charged pions" were to be found instead of acoplanar $(e\mu)$ pairs. Had he followed their "indication", Perl would never have discovered the HL at SLAC.

2.6 — Okun'.

It is interesting to note that all the papers quoted above originate from physicists who have never claimed what is now attributed to them by Perl. The exception is L. Okun', who claimed in a private correspondence with me [1] to be the author of the "first paper in which the idea of the sequential lepton with its own neutrino was first published". Okun''s paper was first published in Russian in 1964, then in English in 1965 [11]. A complete analysis of the 1965 Okun' paper is given in my Review Paper [1]. The main points can be summarised as follows.

Okun' suggested nearly two dozen particles, among which three are possible heavy leptons, but none of which is the HL. Okun' should have introduced a special property (a sort of "leptonic parity" for example) in order to give an exact content to his proposals. Had he done this, he would have realised that three symbols were needed — for example L_{Aa}, L_{Ab}, L_B — in order for people to understand that what he was proposing corresponded to three possible types of heavy leptons, all of which have different properties from the HL (i.e. none of them corresponds to what is now called the "sequential" heavy lepton τ). This proves that the idea of the HL was not a trivial one.

A careful reading of Okun''s paper allows one to understand that his new suggestion was for a heavy lepton carrying the same leptonic number as all the other known leptons. The new heavy lepton proposed by Okun' was heavy also in its neutral component, called L^0.

Nowhere does Okun' say that L^0 could be massless. In Table II of his paper [11], L^0 had to be massive. The charged heavy lepton L^- could either be coupled to the known neutrinos — and in this case could have been produced in neutrino beams — or not be coupled to the known neutrinos but only to L^0. In this case the new heavy lepton could not be short-lived, and in fact could even be stable. It carried the same leptonic number as the other leptons and is different from them by a sort of "leptonic parity" [1]. However, the crucial point is that in Okun"s paper all leptons carry the same leptonic number. It is not a new leptonic number that forbids the production of Okun"s heavy leptons in neutrino interactions, but a sort of "leptonic parity". Okun"s heavy leptons have nothing to do with my work on HL, nor therefore with the third lepton discovered at SLAC in 1975.

3 — Criticisms of the Bologna-CERN-Frascati (BCF) experimental set-up.

This paragraph deals with rumours, consisting mostly of arguments presented to various members of the scientific community, to counter the evidence contained in the papers published by me and my collaborators which clearly establish the priorities in the field of HL [12]. It is not considered dignified to respond to rumours, but for the sake of informing the scientific community which has been exposed to such rumours I have decided to discuss them briefly, as follows. The parts in italic are the rumours.

3.1 — Did more than ten years of work for HL at CERN and Frascati have zero influence on its discovery ?

A set of published papers [12] proves that it was my group who improved, during years of R&D work, electron and muon detection technologies, in order to get the best signature, i.e. acoplanar ($e\mu$) pairs, well above the background. Nevertheless the detractors have said that *these studies did not play any role in the actual 1975 discovery of HL, simply because the SLAC set-up had very weak capabilities for electron and muon identification — so weak that their identification had to be done on a statistical basis.*

This only goes to prove that no-one at SLAC had envisaged, when designing MARK I, the need to search for ($e\mu$) final states, produced in (e^+e^-) annihilation.

From all the evidence we could gather, it appears that:

a) Perl had the idea [6] of looking for ($e\mu$) final states only after reading the first published Frascati results [13] in 1970;

b) he could only use the MARK I, the design of which was already frozen and was very poor for electron and muon identification — which again proves that HL and the technologies needed to detect it were certainly not "trivial" or "common knowledge" at the time;

c) realising from the first result published by the BCF group [13] the importance of this new field of experimental investigation, he strove to get first preliminary evidence of $(e\mu)$ — and it took him a long time — using the MARK I as it was built, despite its very poor capabilities for the task;

d) when he finally obtained preliminary evidence on a statistical basis for $(e\mu)$ production, he was able to convince SLAC to add the muon filters to the MARK I set-up. Once the muon signal was confirmed after the addition of the filters, Perl published the results of the first evidence obtained with MARK I.

3.2 — The invention of the "$(e\mu)$ method".

Another rumour refers to the invention of the "$(e\mu)$ method". As certified by N. Cabibbo [9] in the volume "Lepton physics at CERN and Frascati", this was "Nino's idea". The testimony of Victor Weisskopf [8] and of André Petermann [10] corroborate Cabibbo's statement. And the proof is the PAPLEP (Proton AntiProton into LEpton Pairs) set-up built at CERN in the early sixties — a gigantic apparatus for the standards of that time — in order to simultaneously detect electrons and muons (with extraordinarily high rejection power) produced by time-like photons in $(\bar{p}p)$ annihilation [12]. Let me quote a sentence taken from the final PAPLEP paper [14] : "The electronic signal triggered the thinplate spark chambers (for track reconstruction), the electron detectors (for electron detection) and the range chambers (for muon detection)". Why should the electronic trigger be open for electron and muon detection if not for the search of $(e\mu)$ events? But this search did not produce a publication. This brings me to the next chapter.

3.3 — CERN – Why no HL limit was published.

The construction of the first large solid angle experimental set-up at CERN to search for the $(e\mu)$ signature produced by time-like photons in $(\bar{p}p)$ annihilation cannot be considered because it did not produce any published results in the search for HL.

The reason is that $(\bar{p}p)$ annihilation is not a powerful source of time-like photons, and this is due to the strong, but never previously investigated, electromagnetic form factors of the proton in the time-like region. The depression factor, in terms of time-like photons produced, is 500 times below the point-like values. There is no paper published on HL searches but there are three testimonies (by V.F. Weisskopf [8], N. Cabibbo [9] and A. Petermann [10]) covering the thinking and the experimental aims at that time.

Let me quote another sentence from the final PAPLEP paper [14]: "In practice, the cross-section was shown to be so low that only an upper limit for annihilation into lepton pairs of $\sigma_{p\bar{p} \to l\bar{l}}^{point} \leq 0.54$ nanobarn could be obtained". This is the reason why no limit could be published on the first search for HL using time-like photons from $(\bar{p}p)$. In fact, the proton is far from being a point-like source of time-like photons, due to the strong electromagnetic form factor in the time-like region, never investigated before PAPLEP was set-up. This search established that the source of time-like photons could not be $(\bar{p}p)$, greatly focusing attention on another source of time-like photons, i.e. the (e^+e^-) annihilation. This experiment was very important because it proved the validity of the electron and muon identification technologies (pre-shower for electrons and muon punch-through) implemented at CERN for the first time and now used world-wide. It also allowed my group to have the credibility needed [15] in order to perform a search for HL in the new (e^+e^-) collider being designed and built at Frascati.

3.4 — Frascati – Absence of a magnetic field.

It has been said that *the absence of a magnetic field would have endangered the ability of A. Zichichi to observe HL production.*

If this were true, my collaborators and I would have been unable to:

   i)   produce a mass limit for the HL with such an extraordinarily low background level;

   ii) establish the validity of the $(e \neq \mu)$ leptonic selection rule at 7×10^{-5} level;

   iii) observe for the first time QED radiative acoplanar effects for the electron and the muon final states. Everyone else had measured radiative acollinear effects: never acoplanar;

   iv) perform the highest precision QED tests for electrons and muons (these results are still the best, even after more than two decades).

Notice that the acoplanar $(e\mu)$ signal is accompanied by four neutrinos; the $(\pm)$ particle charge determination is irrelevant and the momenta of the two leptons would be needed if the knowledge of the two-particle $(e\mu)$ invariant mass were important. But this mass has no physical meaning. It is needed only to establish that the number of missing particles is at least two. The Frascati set-up had $(\Delta E / E)$ for electrons and muons discrimination more than sufficient to establish the validity of the correct interpretation for the acoplanar $(e\mu)$ events being generated by $(\mathrm{HL} + \overline{\mathrm{HL}})$ production in (e^+e^-) annihilation, i.e. with at least two missing momenta. In MARK I the magnetic field was already there, it was not added to improve the rejection of hadronic background. On the other hand, as already stated, the muon filters outside MARK I were last-minute but vital changes, applied to a detector which had been designed for other purposes. In fact, when Perl [6] read the first (1970) Frascati paper on the HL [13], the MARK I set-up had already been "frozen". So the existence of a magnetic field in Perl's experiment cannot be used to claim that his detector was meant to have better $(e\mu)$ discrimination than our set-up; rather it proves that his instrument was not designed to detect acoplanar $(e\mu)$ events accompanied by four neutrinos. The magnetic field was unnecessary in the Frascati set-up because the $(e\mu)$ rejection was proven already at CERN to be higher than needed. The central point for the detection of HL was the acoplanar $(e\mu)$ signal: the most powerful background rejection is obtained without a magnetic field. The structure of the detector and the presence of a magnetic field in Perl's experiment are the proof that MARK I was not designed for the study of $(e\mu)$ final states from $HL^{\pm}$ decay.

3.5 — Frascati – Charm Threshold.

Another point in the campaign against the Frascati experiment is the claim that *the production of charm would have destroyed the detection efficiency for acoplanar $(e\mu)$ pairs.*

An unbiased analysis of the Frascati data, published before Perl's observation of acoplanar $(e\mu)$ events, leads to the conclusion that charm production could not have destroyed the $(e\mu)$ signal that we, my collaborators and myself, were searching for. In fact, the hadronic background was very abundant in the ADONE energy range. The hadron production, before ADONE started, was expected to consist of the "tails" of the known vector mesons. It was a remarkable discovery at Frascati that multihadron production exceeded expectations. In fact, during the 1974 London Conference, SLAC reported the result on $\sigma(e^+e^- \rightarrow \text{hadrons})$ starting at 3 GeV, confirming the high rate already observed at Frascati. [On that occasion, I failed to convince Burt Richter that he should not have reported on $\sigma(e^+e^- \rightarrow \text{hadrons})$ but on $\sigma(e^+e^- \rightarrow \text{tracks})$, because it was our detector at Frascati that proved [16, 17] those tracks to be of hadronic nature (pions and kaons).] Charm production adds less than a factor of two to the other three quark couplings (u, d, s), and therefore hadron production above the charm threshold would have been totally mastered by our $(e\mu)$ rejection power. Furthermore, the main decay channels (semileptonic decays) of charmed mesons have several charged mesons in the final state and a soft spectrum for the leptons. Our $(e\mu)$ rejection power would have been higher than needed. Finally, the charm threshold is above τ production, thus giving a further possibility of distinguishing signals in two different energy ranges in case this was needed; all the data we were able to get in the ADONE energy range show that this supplementary safety check would have definitely not been needed. We were able to determine the electromagnetic form factors of the pseudoscalar mesons F_π and F_K; this proves once again the power of our detector. I only quote these points to provide an impartial reader with the correct answers to the rumours circulated world-wide against the Bologna-CERN-Frascati set-up, with the sole purpose of allowing uninformed people to believe that, even if the ADONE energy were above threshold, the "Zichichi set-up" would not have been able to detect the $(e\mu)$ pairs from the HL decays.

3.6 — Calibrations at CERN for the Frascati HL set-up.

The Frascati search for HL had a very strong component at CERN since in the Frascati Laboratory there were no beams of electrons, muons, pions, and K-mesons.

The Frascati set-up was the data-taking part of the experiment, whereas an analogous set-up was running at CERN for calibration purposes. While the Frascati set-up was taking data, the CERN set-up was on continuous calibration runs to check, with beams of known particles ($K\,\pi\,\mu\,e$) and momenta, the expected performance of the Frascati set-up. In fact the BCF group was the only one at Frascati able to prove the hadronic nature [16] of the multitrack events observed by other groups at ADONE, and even distinguish pions from K-mesons [17, 18].

I had spent years of work on calibration runs at CERN to guarantee perfect running conditions for the BCF set-up at Frascati. Calibration runs were a most delicate part of the work and involved spending more time at CERN than at Frascati. It is hard to believe but even the type of rumour *"Zichichi was not permanently but only part-time present in Frascati"* has been used. When I was not at Frascati I was at CERN for the calibration runs. The experiment had two components and the calibration part running at CERN was essential.

4 — Conclusions.

The first time the following reaction

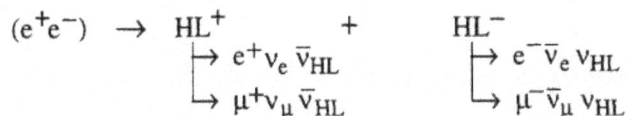

$$(e^+e^-) \;\rightarrow\; HL^+ \qquad\qquad +\qquad\qquad HL^-$$
$$\begin{array}{ll}
\hookrightarrow e^+ \nu_e \bar{\nu}_{HL} & \hookrightarrow e^- \bar{\nu}_e \nu_{HL}\\
\hookrightarrow \mu^+ \nu_\mu \bar{\nu}_{HL} & \hookrightarrow \mu^- \bar{\nu}_\mu \nu_{HL}
\end{array}$$

appeared in the scientific literature was in the INFN proposal [19] for an experiment to be performed with the new (e^+e^-) ADONE collider. The proposal is dated 1967, i.e. one year before Perl's work on the search for new particles in a photoproduction experiment at SLAC, using a single-arm spectrometer [20]. The particles searched for by Perl were stable heavy leptons and have nothing to do with the heavy lepton proposed and searched for by us, using specially designed detectors for the simultaneous identification of electrons and muons. The 1968 paper by Perl is the first evidence that up to that time he had never given any thought to a heavy lepton of the HL type with its own neutrino, and this is corroborated by the set-up he used, MARK I, which had very low ($e\mu$) detection efficiencies. Had M. Perl thought since 1963 — as he now claims [2] — of a heavy lepton like the one my collaborators and I were searching for, first at CERN in the early sixties and later at Frascati, why was his instrument at SLAC so weak in ($e\mu$) identification? This is in contrast with the power of the Bologna-CERN-Frascati detectors implemented at CERN and Frascati, where ($e\mu$) identification was so good that the hadronic background was totally mastered and a series of high-precision QED measurements were performed and final state hadronic processes studied. Had Perl not read the

1970 first Frascati paper, the third lepton would probably have been discovered by somebody else and possibly not at SLAC.

5 —   References.

[1]   *Ten Years of Work for the Third Lepton*
       A. Zichichi.
       La Rivista del Nuovo Cimento, Vol. 3 (1997); on the occasion of the Centenary of the Italian Physical Society (SIF).

[2]   *The discovery of the tau lepton. Part 1: The early history through 1975*
       M. Perl.
       Proceedings of the International Conference on *"The History of Original Ideas and Basic Discoveries in Particle Physics"*, 1994, H.B. Newman and T. Ypsilantis Eds., Plenum Press (1996), Vol. 352, 277.

[3]   *A model of the universal weak interaction*
       E.M. Lipmanov.
       Soviet Physics ZhETF, 43, 893 (1962) and *Soviet Physics JETP, 16, 634 (1963)*.

[4]   *Electron positron colliding beam experiments*
       N. Cabibbo and R. Gatto.
       Physical Review, 124, 1577 (1961).

[5]   *Problems of present-day physics and astronomy*
       Ya. B. Zel'dovich.
       Soviet Physics Uspekhi, 78, 549 (1962) and *Soviet Physics Uspekhi, 5, 931 (1963)*.

[6]   *Are there heavy leptons?*
       K.W. Rothe and A.M. Wolsky.
       Nuclear Physics, B10, 241 (1969).

[7]   *How does the muon differ from the electron?*
       M. Perl.
       Physics Today, 34 (July 1971).

[8]   *The heartbeat of the proton*
       V.F. Weisskopf,
       in *"Lepton Physics at CERN and Frascati"*, N. Cabibbo Ed., 20th Century Physics Series, Vol. 8, World Scientific, 45 (1994).

[9]   *Foreword*
        N. Cabibbo,
        in *"Lepton Physics at CERN and Frascati"*, N. Cabibbo Ed., 20th Century Physics
        Series, Vol. 8, World Scientific, xiii (1994).

[10]   *The roots of the third family*
        A. Petermann,
        in the present volume.

[11]   *On the possible types of elementary particles*
        L. Okun'.
        Soviet Physics ZhETF, 47, 1777 (1964) and *Soviet Physics JETP, 20, 1197 (1965)*.

[12]   *Lepton Physics at CERN and Frascati*
        N. Cabibbo Ed., 20th Century Physics Series, Vol. 8, World Scientific, (1994). This
        volume contains all the references to the HL work performed at CERN and Frascati
        including the reproduction of the original works on the "pre-shower" method to
        improve electron detection and the "muon punch-through" studies to improve muon
        detection.

[13]   *Limits on the electromagnetic production of heavy leptons*
        V. Alles-Borelli, M. Bernardini, D. Bollini, P.L. Brunini, T. Massam, L. Monari,
        F. Palmonari and A. Zichichi.
        Lettere al Nuovo Cimento, 4, 1156 (1970).

[14]   *The leptonic annihilation modes of the proton-antiproton system at 6.8 $(GeV/c)^2$
        timelike four-momentum transfer*
        M. Conversi, T. Massam, Th. Muller and A. Zichichi.
        Nuovo Cimento, 40, 690 (1965).

[15]   *The basic steps which led to the discovery of the heavy lepton τ: a historical record*
        C. Villi.
        Nuovo Cimento, 107 A, 665 (1994); reproduced in the present volume.

[16]   *Proof of hadron production in e^+e^- interactions*
        V. Alles-Borelli, M. Bernardini, D. Bollini, P.L. Brunini, E. Fiorentino, T. Massam,
        L. Monari, F. Palmonari, G. Valenti and A. Zichichi.
        Proceedings of the International Conference on *"Meson Resonances and Related
        Electromagnetic Phenomena"*, Bologna, Italy, 14-16 April 1971 (Editrice Compositori,
        Bologna, 1972), 489.

[17]   *Proof of comparable K-pair and π-pair production from time-like photons of 1.5, 1.6,
       and 1.7 GeV, and determination of the K-meson electromagnetic form factor*
       M. Bernardini, D. Bollini, P.L. Brunini, E. Fiorentino, T. Massam, L. Monari,
       F. Palmonari, F. Rimondi and A. Zichichi.
       Physics Letters, <u>44B</u>, 393 (1973).

[18]   *The time-like electromagnetic form factors of the charged pseudoscalar mesons from
       1.44 to 9.0 GeV^2*
       M. Bernardini, D. Bollini, P.L. Brunini, E. Fiorentino, T. Massam, L. Monari,
       F. Palmonari, F. Rimondi and A. Zichichi.
       Physics Letters, <u>46B</u>, 261 (1973).

[19]   *A proposal to search for leptonic quarks and heavy leptons produced by ADONE*
       M. Bernardini, D. Bollini, E. Fiorentino, F. Mainardi, T. Massam, L. Monari,
       F. Palmonari and A. Zichichi.
       INFN/AE-67/3, 20 March 1967.

[20]   *Search for new particles produced by high energy photons*
       A. Barna, J. Cox, F. Martin, M.L. Perl, T.H. Tan, W.T. Toner, T.F. Zipf and
       E.H. Bellamy.
       Physical Review, <u>173</u>, 1391 (1968).

POSTSCRIPT

by the Editors

By the time this volume was ready for publication, M. Perl's book, *"REFLECTIONS ON EXPERIMENTAL SCIENCE"*, had been published [20th Century Physics Series, Vol. 14, World Scientific (1996)]. The first 344 pages of this book are devoted to the heavy lepton and the first article: *"A memoir on the discovery of the tau lepton and commentaries on early lepton papers"* (pp. 3-29), is (almost) identical to the report: *"The discovery of tau lepton - Part 1: The early history through 1975"*, presented in 1994 by M. Perl at the International Conference on "The History of Original Ideas and Basic Discoveries in Particles Physics" [H.B. Newman and T. Ypsilantis Eds., Vol. 352, 277, Plenum Press (1996)]. This report has been already analysed and commented in the present volume. Hence no further remarks are called for.

On the other hand, we have found very instructive the last article of Perl's book, entitled as the book itself. Especially the last paragraph of this article (p. 537) where Perl writes: *"A final reflection on new ideas. I find it remarkable that a new idea in experimental science is often obvious after someone else gets the new idea"*.

It is a well established fact that the *"new idea"* for a heavy lepton, carrying its own leptonic number and coupled to its own neutrino, as well as the original method of using the acoplanar $(e\mu)$ signal for its detection were both conceived and implemented by A. Zichichi and are of great relevance for *"experimental science"*. This is how it came about that, many years later (after the sixties), others found, all of a sudden, that the *"new idea"* of the heavy lepton and of the $(e\mu)$ method became *"obvious"*.

THE ROOTS OF THE THIRD FAMILY

1 — My recollection.

.... As Weisskopf recalls [2], this proposal by Nino was made before the discovery by Lederman, Schwartz and Steinberger of the two neutrinos $\nu_\mu \neq \nu_e$. The remarkable fact is that Nino did not limit himself to make a suggestion and forget it. He went on discussing with me and his other close friends his ideas on how to search for such a new lepton, emphasizing the technology that he planned to develop which would make the detection of acoplanar $e\mu$ pairs easy. For a new heavy lepton with its leptonic quantum number and with its own neutrino, the best production process was via a time-like photon. Nino decided to take the search for such a heavy lepton as a topic to be seriously investigated. For this search to become effective, an intense source of time-like photons was necessary. ....

....

.... Those were the times of strong interactions and bubble chamber physics. Zichichi succeeded in convincing Weisskopf (CERN DG at that time) to built the first high-intensity, partially separated beam of antiprotons. Unfortunately the time-like structure of the proton was far from being point-like and therefore the $\bar{p}p$ annihilation could not be the right production process for heavy lepton pairs.

It took many years to reach this conclusion and Nino got the green light by Weisskopf to use a new source of time-like photons for the production of heavy lepton (HL) pairs: the new e^+e^- collider being planned at Frascati. ....

....

.... I remember his firm position that the Frascati energy level had to be increased as much as possible. In brief, if it were not for Nino's engagement, the search for a heavy lepton having as signature acoplanar $e\mu$ pairs would never have been started either at CERN or at Frascati.

....

André Petermann

THE SEARCH FOR HEAVY LEPTONS BY A. ZICHICHI
AND HIS COLLABORATORS

A. Zichichi and his group have set out to study $p\bar{p} \rightarrow ee, \mu\mu$ starting 1960. As V. Weisskopf recalls, the underlying aim of this research was to establish $p\bar{p}$ scattering as a source of massive photons from annihilation ($p\bar{p} \rightarrow \gamma^*$) so that $p\bar{p}$ scattering could be used for the search of a new heavy lepton.

In order to achieve its goals the group developed, during several years of research, the preshower technique for separating electrons from hadrons with high degree of certainty (see [1, 2, 3]) and to improve the identification of muons (see [4]). ....
....

In 1967 the group (now called the BCF group) submitted a proposal to the Frascati Laboratory entitled: A Proposal to Search for Leptonic Quarks and Heavy Leptons at ADONE (see [10]). The paper discussed, amongst other topics, a proposal to search for heavy leptons produced by e^+e^- annihilation. These hypothetical heavy leptons $HL^\pm$ were assumed to carry a new lepton quantum number. As a result, they would decay via

$$HL \rightarrow e\nu_L\bar{\nu}_e$$

and

$$HL \rightarrow \mu\nu_L\bar{\nu}_\mu \;.$$

The best way of searching for pair production of HL would be to look for events with nothing but acoplanar e and μ observed in the final state plus missing momentum. The proposed detector was based on the techniques for e / hadron and μ / hadron separation developed by the group in their previous experiments which had reached a π/e suppression of $5 \cdot 10^{-4}$, (see Figs. 1[a, b, c] [2, 3]) for momenta up to 2.5 GeV / c. The π/μ suppression was decreasing from $1.8 \cdot 10^{-2}$ at 1.25 GeV / c to $1.5 \cdot 10^{-3}$ at 2.5 GeV / c (see Fig. 2 and Table I [6]).

....

The group then undertook a massive search for heavy leptons scanning the beam energy range from E = 0.6 to 1.5 GeV, 1.5 GeV being then maximum energy accessible at ADONE (see [14, 15]). ....
.... The envisaged methodology, namely the search for acoplanar $e\mu$ events would have been well suited and the e/π and μ/π separation capabilities of the detector would have allowed to discover the τ. Presumably, the BCF group would have continued to search for the τ in the same way as before, namely by increasing the beam energy of ADONE in steps of 0.1 GeV and collecting at each energy a substantial amount of luminosity. With about 400 nb^{-1} per energy setting no signal would have been observed at $E_{beam} \leq 1.7$ GeV while approximately 10, 19 and 20 signal events from $\tau\bar{\tau}$ production would have been found at $E_{beam} = 1.8$, 1.9 and 2.0 GeV, respectively.

....

Björn H. Wiik and Günter Wolf

A SELECTED SAMPLE OF BASIC POINTS ON THE
SEARCH FOR HL AT CERN AND FRASCATI

....

3.2 — The invention of the "$(e\mu)$ method".

....

.... Let me quote a sentence taken from the final PAPLEP paper [14] : "The electronic signal triggered the thinplate spark chambers (for track reconstruction), the electron detectors (for electron detection) and the range chambers (for muon detection)". Why should the electronic trigger be open for electron and muon detection if not for the search of $(e\mu)$ events? But this search did not produce a publication. This brings me to the next chapter.

3.3 — CERN – Why no HL limit was published.

....

The reason is that $(\bar{p}p)$ annihilation is not a powerful source of time-like photons, and this is due to the strong, but never previously investigated, electromagnetic form factors of the proton in the time-like region. The depression factor, in terms of time-like photons produced, is 500 times below the point-like values. There is no paper published on HL searches but there are three testimonies (by V.F. Weisskopf [8], N. Cabibbo [9] and A. Petermann [10]) covering the thinking and the experimental aims at that time.

Let me quote another sentence from the final PAPLEP paper [14]: "In practice, the cross-section was shown to be so low that only an upper limit for annihilation into lepton pairs of $\sigma_{p\bar{p} \to l\bar{l}}^{point} \leq 0.54$ nanobarn could be obtained". This is the reason why no limit could be published on the first search for HL using time-like photons from $(\bar{p}p)$. ....

....

3.4 — Frascati – Absence of a magnetic field.

....

Notice that the acoplanar $(e\mu)$ signal is accompanied by four neutrinos; the $(\pm)$ particle charge determination is irrelevant and the momenta of the two leptons would be needed if the knowledge of the two-particle $(e\mu)$ invariant mass were important. But this mass has no physical meaning. It is needed only to establish that the number of missing particles is at least two. The Frascati set-up had $(\Delta E / E)$ for electrons and muons discrimination more than sufficient to establish the validity of the correct interpretation for the acoplanar $(e\mu)$ events being generated by $(HL + \overline{HL})$ production in (e^+e^-) annihilation, i.e. with at least two missing momenta. In MARK I the magnetic field was already there, it was not added to improve the rejection of hadronic background. On the other hand, as already stated, the muon filters outside MARK I were last-minute but vital changes, applied to a detector which had been designed for other purposes. In fact, when Perl [6] read the first (1970) Frascati paper on the HL [13], the MARK I set-up had already been "frozen". So the existence of a magnetic field in Perl's experiment cannot be used to claim that his detector was meant to have better $(e\mu)$ discrimination than our set-up; rather it proves that his instrument was not designed to detect acoplanar $(e\mu)$ events accompanied by four neutrinos. The magnetic field was unnecessary in the Frascati set-up because the $(e\mu)$ rejection was proven already at CERN to be higher than needed. The central point for the detection of HL was the acoplanar $(e\mu)$ signal: the most powerful background rejection is obtained without a magnetic field. The structure of the detector and the presence of a magnetic field in Perl's experiment are the proof that MARK I was not designed for the study of $(e\mu)$ final states from $HL^{\pm}$ decay.

3.5 — Frascati – Charm Threshold.

....

.... Charm production adds less than a factor of two to the other three quark couplings (u, d, s), and therefore hadron production above the charm threshold would have been totally mastered by our $(e\mu)$ rejection power. Furthermore, the main decay channels (semileptonic decays) of charmed mesons have several charged mesons in the final state and a soft spectrum for the leptons. Our $(e\mu)$ rejection power would have been higher than needed. Finally, the charm threshold is above τ production, thus giving a further possibility of distinguishing signals in two different energy ranges in case this was needed; all the data we were able to get in the ADONE energy range show that this supplementary safety check would have definitely not been needed. We were able to determine the electromagnetic form factors of the pseudoscalar mesons F_{π} and F_K; this proves once again the power of our detector. ....

3.6 — Calibrations at CERN for the Frascati HL set-up.

The Frascati search for HL had a very strong component at CERN since in the Frascati Laboratory there were no beams of electrons, muons, pions, and K-mesons.

The Frascati set-up was the data-taking part of the experiment, whereas an analogous set-up was running at CERN for calibration purposes. While the Frascati set-up was taking data, the CERN set-up was on continuous calibration runs to check, with beams of known particles $(K \pi \mu e)$ and momenta, the expected performance of the Frascati set-up. In fact the BCF group was the only one at Frascati able to prove the hadronic nature [16] of the multitrack events observed by other groups at ADONE, and even distinguish pions from K-mesons [17, 18].

....

Antonino Zichichi

www.ingramcontent.com/pod-product-compliance
Lightning Source LLC
Chambersburg PA
CBHW081123050426
42445CB00002BA/2